Praise for **WHAT THE LUCK?**

"Gary Smith's *What the Luck?* alerts us to many subtle and unappreciated consequences of one of life's great truths: it has its ups and downs."
—GEORGE AKERLOF, Georgetown University,
Nobel Laureate in Economics 2001

"*What the Luck?* is a fabulous, funny, and completely accessible explanation of the pervasive misreporting and misunderstanding of what statistics really mean in common discourse. If you are interested in what is really going on in sports, gambling, genetic inheritance, and everything else that is misrepresented statistically, you will love this book. I wouldn't wait for the TV miniseries, I'd take *What the Luck?* to bed now."
—BRUCE CHAPMAN, Professor of Economics,
Australian National University

"This book deeply enriches readers' understanding of the ubiquitous role of chance in everyday life. The book is so wonderful that I predict the author's next book will be less wonderful. Why the pessimistic prediction? This book will painlessly—even enjoyably—teach you why."
—MICHAEL MURRAY, Charles Franklin Phillips Professor, Bates College

"*What the Luck?* is a humorous, entertaining book citing real-life examples, from areas as diverse as gambling, sports, scholastic achievement, medicine and the stock market to explain how randomness and 'luck' are far more prevalent in our daily lives than we may realize. Reading this book also helped remind me that regression is often a more meaningful driver of stock (and bond) performance than the underlying business fundamentals. This is an important lesson for any serious investor."
—MIKE SCHIMMEL, Portfolio Manager, Kayne Anderson Capital Advisors

"*What the Luck?* is a must-read for those in the healthcare profession. We are constantly inundated with research on the latest and greatest therapy and understanding regression toward the mean can help us better interpret and apply this research to the daily care of our patients."
—ROBERT SALLIS, MD, Director of Sports Medicine, Kaiser Permanente
Medical Center, Clinical Professor of Family Medicine at UC Riverside
School of Medicine, Past-President of the American College of Sports
Medicine (ACSM), Chair of Exercise is Medicine, a joint initiative of
ACSM and the American Medical Association

"In clear, entertaining prose and the use of telling, useful, and even charming examples, Smith dissects one of the most fundamental principles of how the world works—and how our intuitions often fail to catch on. Anyone who wants to think more clearly and act more rationally will profit from reading this book." —TOM GILOVICH, author of *How We Know What Isn't So* and Professor at Cornell University

"There is an infectious clarity of statistical reasoning in Gary Smith's work. His friendly, logical, systematic writing entertains and gives a confidence of membership in an inner circle as brilliant as Smith himself." —ERIC ENGBERG, Data Scientist and Software Engineer, Wells Fargo

"People often underestimate the impact of luck in their lives. *What the Luck?* is eminently readable and entertaining, filled with colorful examples." —SEBASTIAN THOMAS, Director, Head of US Technology Research, Allianz Global Investors

"The beauty of this book is it sheds light on the need for humility when one experiences good fortune, and the importance of hope after misfortune. This nuanced understanding will help readers make better decisions in all realms of their lives." —JONATHAN ABELSON, MD, radiation oncologist

"Regression toward the mean is the key to the universe. Of course, learning about this is both a blessing and a curse. Once people have digested this book, they will absolutely see regression everywhere and understand its effect, but they will also be driven crazy, as you undoubtedly are, when they hear all of the unsupported and sometimes absurd explanations people cling to in order to make sense of it." —JAY CORDES, Data Scientist

"Humans are prone to search and 'find' causal drivers of the events that shape our lives. In reality, we are impacted by chance more often than we think. Professor Smith uses simple reasoning and vivid examples to help us decipher truth from fiction, thereby helping us to make better decisions." –BRYAN WHITE, Founder, Sahsen Ventures

"Smith uses a wide variety of real-life examples to illustrate why conventional wisdom often fails to acknowledge that one of the most important ingredients is luck." —KARL J. MEYER, Director of Strategy, Hewlett Packard Enterprise

"As a lifelong sports fan I've always been puzzled by a phenomenon in which a team wins a championship and, after failing to repeat the next season, everyone from fans to management looks for what went wrong and begins trading players and changing strategies, thereby dooming the team to do even worse the season after. Gary Smith has solved this puzzle, and many more in all walks of life, through the concept of regression to the mean, one of the most powerful and least understood factors in how things turn out in life. You will not look at the world the same after reading this illuminating book."

—MICHAEL SHERMER, publisher of *Skeptic* magazine, columnist for *Scientific American*, and author of *The Moral Arc and Why People Believe Weird Things*

"Absorbing Gary Smith's message will help us avoid common but costly mistakes like putting too much of our wealth in investments that have little probability of out-performing the market, or expecting ourselves (or others) to perform in ways we (or they) can't, or feeling guilty for what happens that isn't really our fault. Lots of people everywhere can benefit from the principle he so clearly explains, and many of the concrete examples he uses to illustrate it will strike home." —BENJAMIN M. FRIEDMAN, William Joseph Maier Professor of Political Economy, Harvard University, and author, *The Moral Consequences of Economic Growth*

"Read this book. Then give it to your family and friends. There is no other single idea that will better improve your understanding of the world, and judgment of the future, than regression to the mean. Drawing on education, health, politics, business, and sports, Smith shows us how others have gotten it wrong and how you can get it right."

—CADE MASSEY, Professor, Wharton School of Business

"*What the Luck?* is a tremendously entertaining and revealing read. A quick and engrossing piece of work, it is a must read for those who approach the world with educated insight!" —SIMEON NESTOROV, CFA, Managing Director, Berkeley Square, Inc.

"Vagaries of chance are part of our lives, whether we like it or not. *What the Luck?* presents serious stuff in an eminently readable and entertaining manner. Using colorful examples, it teaches humility for good fortune and hope after misfortune. A wonderful read!"

—CRISTIAN CALUDE, Professor of Applied Science, University of Auckland, and Giuseppe Longo, Centre Cavaillès, CNRS et Ecole Normale Supérieure, and the Department of Integrative Physiology and Pathobiology, Tufts University School of Medicine

"Gary Smith has another winner! His ability to combine entertaining writing with meaningful analysis should put him at the top of every thinking person's reading list." —WOODY STUDENMUND, Laurence de Rycke Professor of Economics, Occidental College

"Decision makers everywhere should read it to avoid making the mistakes of their predecessors." —ANDREW GELMAN, Professor of Statistics and Political Science, Director of the Applied Statistics Center at Columbia University

"Smith uses entertaining and intuitive examples to show how regression to the mean explains patterns in education, business, and medicine." —ANITA ARORA, MD, MBA, RWJF Clinical Scholar at Yale University

"Smith provides a fascinating and accessible overview of regression toward the mean in sports and other domains. If you play fantasy sports, you should go get a copy of this book at once (while hoping your competitors have not done likewise)." —ALAN REIFMAN, author of *Hot Hand: The Statistics Behind Sports' Greatest Streaks*

"If you can combine the insightful lessons of this book with equanimity in decision-making, your foresight may become remarkable—and it won't be due to luck." —MICHAEL SOLOMON, Partner, Leonard Green & Partners, Private Equity

"Few statistical concepts are as important to understand today as regression to the mean. Through fascinating tales, *What the Luck?* provides a panoply of examples of this essential phenomenon in sports, business, life, and more. I heartily recommend it." —PHIL SIMON, award-winning author of *Message Not Received: Why Business Communication Is Broken and How to Fix It*, lecturer at Arizona State University's W. P. Carey School of Business

"After reading *What the Luck?*, you will appreciate how to separate the sense from the nonsense when it comes to making decisions about your health, your money, your test scores, or your favorite sports team. Written with accuracy and humor, I highly recommend it." —ARTHUR BENJAMIN, Professor of Mathematics, Harvey Mudd College author of *The Magic of Math: Solving for x and Figuring Out Why*

WHAT THE LUCK?

ALSO BY GARY SMITH

Standard Deviations:
Flawed Assumptions, Tortured Data,
and Other Ways to Lie with Statistics

What the Luck?

The Surprising Role of Chance in Our Everyday Lives

Gary Smith

The Overlook Press
New York, NY

This edition first published in hardcover in the United States in 2016
The Overlook Press, Peter Mayer Publishers, Inc.

NEW YORK
141 Wooster Street
New York, NY 10012
www.overlookpress.com
For bulk and special sales, please contact sales@overlookny.com,
or write us at the above address.

Cataloging-in-Publication Data is available from the Library of Congress

Book design and typeformatting by Bernard Schleifer
Manufactured in the United States of America
FIRST EDITION
ISBN 978-1-4683-1375-8
2 4 6 8 10 9 7 5 3 1

To my wife Margaret
and my children Cameron, Chaska,
Claire, Cory, Jo, and Josh

There are few statistical facts more interesting than regression to the mean for two reasons. First, people encounter it almost every day of their lives. Second, almost nobody understands it. The coupling of these two reasons makes regression to the mean one of the most fundamental sources of error in human judgment.

—ANONYMOUS

Contents

I. OVERREACTION

1

The Law of Small Numbers

ELISHA ARCHIBALD MANNING III SOUNDS MORE LIKE A MEMBER OF the English royalty than an American football player, but "Archie" Manning was indeed a terrific football player. Growing up in a tiny town in Mississippi, he starred in baseball, basketball, football, and track in high school and was drafted four times by Major League Baseball teams. Archie decided to play football at the University of Mississippi and had a legendary career there despite the team's otherwise modest talent. One year, he was third in the voting for the Heisman Trophy, honoring the nation's outstanding college football player; another year, he was fourth. The speed limit on the Mississippi campus is now 18 miles per hour, honoring Archie's uniform number.

He was the second player chosen in the National Football League (NFL) draft in 1971. Unfortunately, it was by the New Orleans Saints, a team so awful that they were nicknamed the Aints. Their unhappy fans started the hilarious tradition of wearing paper bags over their heads so their friends wouldn't know that they bought tickets to watch the Aints lose yet another game.

Archie married the college homecoming queen and they had three sons—Cooper, Peyton, and Eli. Cooper's football career was cut short by a spinal problem. Eli plays quarterback for the New York Giants and twice led them to Super Bowl victories. Peyton was also an NFL quarterback and wore number 18, just like his dad. When Peyton retired after quarterbacking the Denver Broncos to a Super Bowl victory in 2016, he had been selected five times as the league's Most Valuable

Player and held NFL career records for the most passing yards, touchdown passes, and wins.

The week before the start of the 2014 NFL season, I listened to several ESPN commentators predict how Peyton would perform in 2014. It was a great example of the law of small numbers.

Elite quarterbacks have many skills, including reading defenses, finding open receivers, and throwing accurate passes. There is a complex NFL rating formula for evaluating quarterback performances based on the percentage of passes completed, average yards gained per throw, percentage of passes that were touchdowns, and percentage of passes that were intercepted.

Even as great a quarterback as Peyton Manning has his ups and downs. Figure 1 shows Peyton's quarterback rating for each regular-season game in 2013. Peyton was a veteran quarterback, headed for the Hall of Fame. His ability was much more stable than the game-to-game swings in his quarterback rating. The fluctuations in his quarterback rating from one game to the next demonstrate how athletic performances are affected by luck—good luck at times, bad luck other times. Sometimes, a defender raises a hand at just the right moment and deflects a pass; sometimes he doesn't. Sometimes, a receiver drops a well-thrown pass; sometimes he catches a poorly thrown pass. Sometimes, a fumbled ball is recovered by the team that fumbled; sometimes, it isn't. Sometimes, the official throws a flag; sometimes, he doesn't. The commonplace refrain, "On any given Sunday, any team can beat any other team," is based on the reality that there is a lot of luck in football games. Yet, coaches, players, fans, and ESPN commentators do not understand the implication. Because performances are affected by luck, extreme performances typically are followed by performances that are less extreme.

In Peyton Manning's case, it is not surprising that after an unusually high quarterback rating in the fourth game of the 2013 season, he did not do as well the next game, and that after a low quarterback rating in game 11, he did better the next game. Figure 1 shows that other great games tended to be followed by games that were not as great, while bad games (by Peyton's lofty standards) tended to be followed by better games.

Figure 1
Peyton Manning's Quarterback Rating, 2013 Regular Season

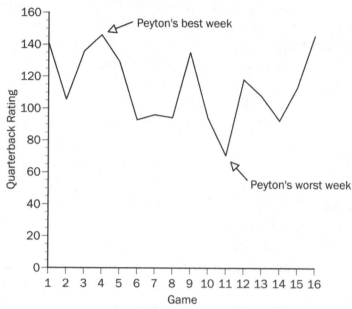

That is the nature of the beast called luck. Peyton's ability did not gyrate wildly game to game, but his luck fluctuated, causing his quarterback rating to zig and zag. When Peyton had good fortune one game, he was unlikely to have as much good fortune the next game. If we do not consider the importance of luck, we might expect that every great game will be followed by an equally great game. When it doesn't happen and his performance dips, we might speculate that he was lazy or perhaps jinxed by success, instead of recognizing that his luck simply changed.

Psychologist Daniel Kahneman was awarded a Nobel prize in economics for his work with Amos Tversky in identifying and documenting ways in which humans differ from the completely rational automatons assumed by classical economic models. (Tversky was deceased or he would have received the prize, too.) One of these human foibles is a fallacious reasoning Kahneman and Tversky call "the law of small numbers." An example of this error is when we see someone correctly predict three out of four football games, presidential elections, or stock market movements, and we assume that this person is generally right seventy-five percent of the time. If so, we are overreacting to very limited data, making generalizations when there is no persuasive reason for doing so. It is like seeing a coin land heads three times in four tosses and concluding that heads come up 75 percent of the time. The reason we don't draw this hasty conclusion about coins is that we know the coin has two sides and believe that each is equally likely. In sports, politics, and the stock market, however, there is no coin to inspect and it is tempting to overreact to a small number of successes or failures.

It is a law-of-small-numbers fallacy to see a great athletic performance and assume that it is an accurate measure of the athlete's ability. Exceptional performances typically involve good fortune—which means that a remarkable performance usually exaggerates the athlete's ability. Not only that, good fortune cannot be counted on indefinitely, so great performances are typically followed by not-so-great performances. Not necessarily bad performances, just performances that are less exceptional.

In the same way, below-par performances usually involve bad luck and are followed by better performances. It is as if there is a mediocrity magnet in that extraordinary performances—good or bad—are typically followed by less remarkable performances. Statisticians call this mediocrity magnet "regression to the mean." The concept is simple, but powerful. The key is recognizing it. As the epigraph to this book says:

> There are few statistical facts more interesting than regression to the mean for two reasons. First, people encounter it almost every day of their lives. Second, almost nobody understands it. The coupling of these two reasons makes regression to the mean one of the most fundamental sources of error in human judgment.

The reasons why performances bounce around ability are as varied as the scenarios. A student might get an unusually high test score because some guesses turned out to be correct. A healthy person might get a worrisome medical test result because the equipment isn't completely clean. A study of a new medical treatment might show spectacular success because the people treated happened to be unusually healthy. A company might have unusually high earnings because of a favorable news story. A job candidate might ace a job interview because she happened to have been asked questions she spent a lot of time thinking about ahead of time. A quarterback might throw an interception because the intended receiver slips and falls.

Let's apply this reasoning to Peyton Manning's quarterback rating. Look again at Figure 1. It would be a law-of-small-numbers fallacy to see Peyton's 146 rating in game 4 and assume that it is an accurate assessment of his ability. Considering his performances throughout his long career, Peyton surely had good fortune in game 4, and it is no surprise that he did not do as well in game 5. In game 11, in contrast, Peyton's 70 rating no doubt involved some bad luck, and he did better in the following game. The more extreme the luck—good or bad—the more likely it is to be followed by less extreme luck. That's why exceptional performances—good or bad—tend to regress to the mean.

This reasoning also applies to Peyton's quarterback rating for the season as a whole. Figure 2 shows Peyton's quarterback rating each year from 1998 through 2013 (with the exception of 2011, which he missed because of neck surgery and cervical fusion surgery).

Figure 2
Peyton Manning's Quarterback Rating, 1998-2013

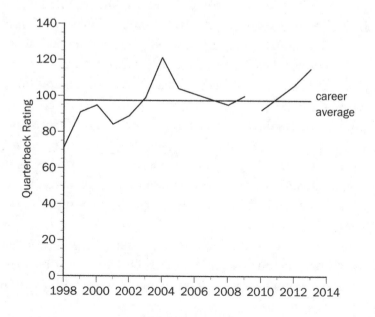

The first few years of his career, Peyton's ability may have been improving as he made the transition from college football to the NFL. However, after those first few years, his ability surely did not fluctuate as much as his performance. Most of the variation in Peyton's quarterback rating week to week and season to season were not due to zigs and zags in his ability, but to swings in his fortune and misfortune. It would be a law-of-small-numbers fallacy to look at Peyton's 115.1 quarterback rating in 2013 and conclude that it is an accurate assessment of his ability, yet that is exactly what the ESPN commentators did.

Peyton had an incredible year in 2013, one of the best years in his career. He threw 55 touchdown passes with only 10 interceptions. The next closest in touchdowns was Drew Brees with 39. Peyton's 115.1 quarterback rating was the highest rating for any quarterback with more than 320 passes (20 per game). The next closest were Philip Rivers (105.5) and Drew Brees (104.7).

Looking forward to the 2014 season, the ESPN commentators were seduced by the law of small numbers. They talked about the 2013 season as if that was pretty much all that mattered for the forthcoming 2014 season. They talked about Peyton Manning's pass receivers, the team's running backs, and the offensive line. No one said a word about how Peyton might have been lucky in 2013.

They assumed that Peyton would do just about as well in 2014 as in 2013. They predicted he would throw 48 touchdown passes and 12 interceptions and, once again, be the top NFL quarterback by a wide margin. They predicted that Peyton would rack up 368 fantasy points in Fantasy Football, well above their predictions for second and third place: Aaron Rodgers (347 points) and Drew Brees (329 points).

The commentators were overreacting to Peyton's 2013 stats and ignoring the likely pull of the mediocrity magnet in 2014. They should have looked at his entire career and considered the possibility that Peyton had good luck in 2013, because the more he benefited from good luck, the more likely it is that he will not have as much good luck in 2014.

As I listened to the praise from these commentators, I thought to myself that since Peyton Manning, at age 37, had one of his best years ever in 2013, good luck must have had a lot to do with it. So, I posted a blog entry before the start of the 2014 season titled, "Peyton Manning is Likely to Regress to the Mean." I ended the post with this prediction:

> Peyton Manning's phenomenal 2013 season surely bene-fited from more good luck than bad. Defensive players slip-ping, offensive players not slipping. Defensive players making bad guesses, offensive players making good guesses. Fumbles lost and recovered. Passes caught and dropped. Holding penalties called and not called. The list is very long. Luck—good and bad—is why the best team doesn't win every game, why player stats go up and down from one game to the next.
>
> Manning is a Hall-of-Fame quarterback, but 2013 was not a below-average season for him. Manning is surely not as good as he seemed last year, and almost certainly will not do as well this year. You can take that to the bank.

I was right. Not because I know a lot about football, but because I know something about regression to the mean.

Yep, Peyton regressed in 2014. Instead of the predicted 48 touch-downs with only 12 interceptions, he had 39 touchdowns and 15 in-terceptions. Instead of leading the League with 368 fantasy points, Peyton finished fourth with 307 points, behind Aaron Rodgers (342), Andrew Luck (336), and Russell Wilson (312). Peyton's quarterback rating was 101.5 and he finished fourth, well behind Tony Romo (113.2), Aaron Rodgers (112.2), and Ben Roethlisberger (103.3).

Peyton didn't have a bad year. He was still one of the top quarter-backs. But he didn't have as much good luck in 2014 as the year before, and he didn't finish first. Peyton Manning is human and, like other humans, is susceptible to regression to the mean.

The law of small numbers is a fallacy to avoid. Regression to the mean is a reality to recognize. We should not overreact to limited data and we should not be surprised by the mediocrity magnet. It is true of athletic performances, test scores, medical studies, business profits, job interviews, romance, and much, much more.

II. INHERITED TRAITS

2

The Father of Regression

TALL PARENTS TEND TO HAVE TALL CHILDREN, AND SHORT PARENTS generally have short children but, even after adjusting for gender differences, brothers and sisters are not all the same height. I am 6' 4" and my brother is 6' 1". My sisters are 5' 8" and 5' 11". We inherited tall genes from our parents, but clearly there is more to it than genes.

In the late 1800s Francis Galton made the first systematic study of the relationship between the heights of parents and their children. Galton was a child prodigy and wrote hundreds of papers and books on topics as varied as anthropology, geography, meteorology, psychology, biology, and criminology. In his forties, he was inspired by his half-cousin Charles Darwin's revolutionary book, *The Origin of Species*, to begin his own study of inherited traits; indeed, Galton coined the phrase "nature versus nurture" and pioneered the use of twin studies and adoption studies to estimate the relative importance of nature and nurture.

In one of his studies of inherited traits, Galton weighed thousands of sweet pea seeds to the nearest hundredth of a grain (one grain is 0.00228571429 ounces) and put the seeds in seven weight categories. He then gave seven friends ten seeds in each weight category, along with very detailed planting instructions intended to ensure uniformity. For example, each friend was told to prepare seven planting beds in parallel rows (one for each weight category), with each bed 1.5 feet wide and 5 feet long. Ten 1-inch-deep holes were to be poked in the soil at uniform intervals in each bed, with one seed placed in each hole.

Although Galton had separated the seeds by weights, he found that there was an extremely close relationship between weight and diameter and he chose to present his results by calculating the diameters of the parent seeds and their offspring seeds. Figure 1 shows the average diameter of the seeds in each parent grouping and of their offspring. If each parent grouping had offspring with the same average diameter as their parents, the fitted line would be a 45-degree line going through the origin. The actual slope is 0.34, which means that a parent group with a diameter that is one-hundredth of an inch above (or below) average had offspring that averaged only 0.34 hundredths of an inch above (or below) average.

There is a hereditary component in that larger parent seeds tended to produce larger offspring; however, there is also luck. The largest parent seeds are likely to have had positive environmental influences and the smallest parent seeds are likely to have had negative environmental influences. Environmental influences are not passed on to the offspring, so the offspring seeds are closer to the mean than were their parents. Galton called this pattern "regression" (from a Latin root meaning "going back").

Galton drew lines (like the line in Figure 1) that seemed, to his trained eye, to fit the data as well as possible. This eyeballing isn't scientific since closeness may be in the eye of the beholder. It would be better to have a mathematical formula that could be used to draw the line without having to worry about a person's eyesight and judgment.

Galton's colleague Karl Pearson developed a formula. A reasonable definition of "best fit" is the line that is, overall, as close to the data points as any straight line could possibly be. If we are trying to predict the size of the offspring from the parent, as in Figure 1, it makes sense to look at the vertical distance of each point from the line. It also makes sense to look at the squared distance since large prediction errors are more worrisome than small errors. Pearson worked out the mathematics for determining the line that minimizes the average squared distance of the points from the line. This is now called the "least squares line." However, recognizing Galton's role in collaborating with and inspiring Pearson, it is also called the regression line.

Figure 1
Diameters of Parent and Offspring Sweet Pea Seeds

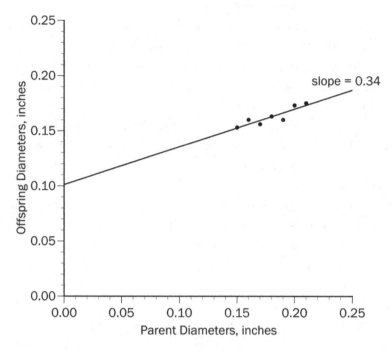

Was the regression that Galton found in sweet pea seeds true of humans, too? Galton couldn't do experiments with humans the way he could with sweet peas so, instead, he gathered data on the heights of hundreds of parents and their adult children. He multiplied each mother's height by 1.08 because the men were, on average, eight percent taller than the women. He then averaged the heights of each mother and father to obtain a mid-parent height. The children and parents turned out to have the same average height: 68.2 inches.

As with the sweet-pea study, he grouped the parents into categories (64 to 65 inches, 65 to 66 inches, and so on) and calculated the median height of the parents and the adult children for each parent category. Figure 2 shows his results, along with a 45-degree line. The positive relationship is due to heredity, but it is an imperfect relationship because of chance factors—what we call luck.

If the parents and children in each category had the same average height, the points would lie on the 45-degree line. Points above the

45-degree line are cases where the children were, on average, taller than their parents; points below the line are cases where the children are shorter than their parents.

Yes, tall parents tend to have tall children, while short parents have short children. However, the points for tall parents are below the 45-degree line because their children are generally not as tall as the parents. The points for short parents are above the line because their children tend to be not as short as the parents.

These are self-reported heights and the data with the question mark in Figure 2 may reflect the seductive appeal of being six-feet tall. Just like $10 sounds much more expensive than $9.99, so being six feet tall sounds much taller than being 5-foot, 11-inches. In Galton's data for parents with a 6-foot mid-parent height, two children were reported to be 5-foot, 11-inches tall; seven were said to be six-feet tall; and two were reported to be 6-foot, 1-inch. The 6-foot blip is probably wishful thinking.

Figure 2
Children Are Closer to the Mean than Are Their Parents

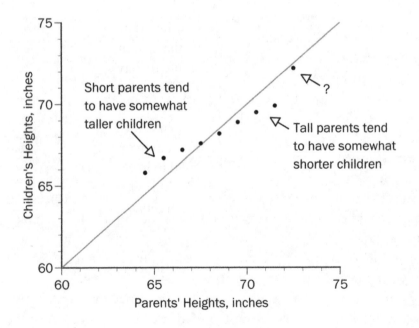

Figure 3
Children's Heights Regress to the Mean

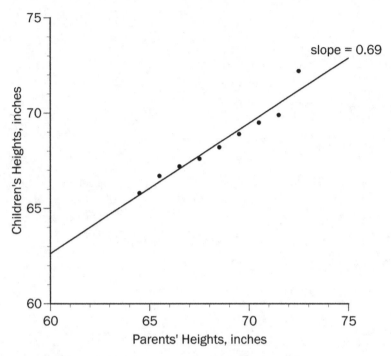

Figure 3 shows that the regression line that best fits Galton's data has a slope of 0.69, which means that parents whose height is one inch above (or below) average tend to have children whose heights are only 0.69 inches above (or below) average.

Regression goes the other way, too, from children to their parents. Galton grouped the children into height categories, from shortest to tallest, and calculated the median mid-parent height in each category. Figure 4 shows that tall children tend to have not-so-tall parents while short children tend to have not-so-short parents.

Regression goes both ways, from parents to their children and from children to their parents, because the explanation is entirely statistical, not causal—for instance, a suspicion that a child's true father is not the husband but another, somewhat shorter man.

Figure 4
Parents Are Closer to the Mean than Are Their Children

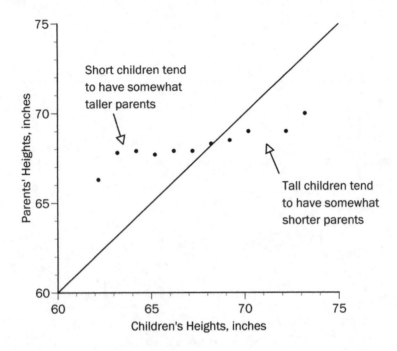

Interpreting Regression

Galton noted that there is a positive correlation between parental and offspring heights due to genetic influences, and he also listed several reasons why the relationship is imperfect:

> Stature is not a simple element, but a sum of the accumulated lengths or thicknesses of more than a hundred bodily parts, each so distinct from the rest as to have earned a name by which it can be specified. The list of them includes about fifty separate bones, situated in the skull, the spine, the pelvis, the two legs, and the two ankles and feet. The bones in both the lower limbs are counted, because it is the average length of these two limbs that contributes to the general stature. The cartilages interposed between the bones, two at each joint, are rather more numerous than the bones

themselves. The fleshy parts of the scalp of the head and of the soles of the feet conclude the list. Account should also be taken of the shape and set of many of the bones which conduce to a more or less arched instep, straight back, or high head.

However, Galton failed to recognize the implications of the chance factors he enumerated. Unusually tall parents are likely to have had positive luck, so their observed heights generally overstate the genetic factors that they inherited from their parents and pass on to their children. Their children are unlikely to have as much positive luck as their parents, and will consequently tend to be shorter than their parents. Similarly, unusually short parents most likely had negative luck, so their children will, on average, have better luck and be taller than their parents. It is as if there is a mediocrity magnet that draws children towards the mean.

Instead of recognizing that chance factors—positive or negative—are not passed on from parent to child, Galton offered a speculative (and incorrect) explanation of regression:

> The child inherits partly from his parents, partly from his ancestry. Speaking generally, the further his genealogy goes back, the more numerous and varied will his ancestry become, until they cease to differ from any equally numerous sample taken at haphazard from the race at large. Their mean stature will then be the same as that of the race; in other words, it will be mediocre.

Galton erroneously believed that the hereditary component of a person's height depends not only on his or her immediate parents but, in addition, on the grandparents, great grandparents, and other ancestors back to the beginning of the human race. Since everyone shares the same distant ancestors, the gene pool from which everyone draws has less variation than one's immediate parents. Therefore, he reasoned, the children are more similar than one might think from merely observing the parents. Here, Galton was wrong.

The inherited traits that parents pass on to their children come from the parents' parents, true enough, but the grandparents have no separate influence beyond what they already contributed to the parents. Regression is a statistical phenomenon that is caused by the fleeting nature of luck, not by the traits of distant ancestors. This is why the mediocrity magnet goes the other way, too. The parents of unusually tall or short children are generally less extreme that their children.

This observation by Galton is closer to the mark:

> The number of individuals in a population who differ little from mediocrity is so preponderant that it is more frequently the case that an exceptional man is the somewhat exceptional son of rather mediocre parents than the average son of very exceptional parents.

A person who is 6-feet, 6-inches tall might have 6-foot genes and experienced positive environmental influences (positive luck) or might have 7-foot genes and had negative environmental factors (negative luck). The former is more likely, simply because there are many more people with 6-foot genes than with 7-foot genes. Thus, the observed heights of very tall parents usually overstate their genetic heights and, as a consequence, the average height of their children.

Galton titled his research report, "Regression Towards Mediocrity in Hereditary Stature," which might encourage the erroneous conclusion that heights are regressing to mediocrity, with everyone eventually being the same height. If, as in Galton's study, parents whose height is 1 inch above (or below) average tend to have children whose heights are only 0.69 inches above (or below) average, won't the next generation being even closer to average, and the generation after that even closer? Nope. Regression toward the mean does not imply that everyone will soon be the same height any more than it implies that every NFL quarterback will soon be equally mediocre.

There have been thousands of generations of humans. If human heights were converging to a uniform mediocrity, it would have happened by now. Despite the title of his study, Galton observed that there is just as much variation in the heights of children as in the heights of their parents. He explained this by envisioning an almost magical balance in nature that allows the random component of children's height to just offset the generational regression, thereby maintaining a natural equilibrium:

> How is it, I ask, that in each successive generation there proves to be the same number of men per thousand, who range between any limits of stature we please to specify, although the tall men are rarely descended from equally tall parents, or the short men from equally short? . . . The answer is that the process comprises two opposite sets of actions, one concentrative and the other dispersive, and of such a character that they necessarily neutralize one another, and fall into a state of stable equilibrium . . . The stability of the balance between the opposed tendencies is due to the regression being proportionate to the deviation. It acts like a spring against a weight; the spring stretches until its resilient force balances the weight, then the two forces of spring and weight are in stable equilibrium; for if the weight be lifted by the hand, it will obviously fall down again when the hand is withdrawn and, if it be depressed by the hand, the resilience of the spring will be thereby increased, so that the weight will rise when the hand is withdrawn.

That is a wonderful image, but there is no need for such imaginative speculation. We now know that regression is a purely statistical consequence of the fact that heights have a random component. Regression works in both directions because it reflects nothing more than random fluctuations. Tall parents tend to have somewhat shorter children, and tall children tend to have somewhat shorter parents. That's it. There is no elaborate set of springs and weights created by a benevolent nature to maintain a magical equilibrium.

I am 6-foot, 4-inches tall. It would be a law-of-small-numbers fallacy to guess that my parents, siblings, and children are 6-feet, 4-inches, too. My parents were tall, my siblings are tall, and my children are tall. There are, no doubt, tall genes in my family. However, I also had positive luck. I am taller than my parents, my siblings, and my adult children.

3

Choose Your Parents Carefully

EIGHT AND WEIGHT ARE AMONG MANY TRAITS THAT ARE INFLU-
enced, but not completely determined, by genes that pass
from one generation to the next. What we see (for example,
a woman's height) is not necessarily what she got (her genes). Whenever this is the case, the logic of regression tells us that abnormal parents generally have less abnormal children, and abnormal children
typically have less abnormal parents.

Ever since Galton, a great deal of research on inherited traits has
focused on intelligence. Galton didn't have data (like scores on intelligence tests) that he could use to compare parents' and children's intelligence the way that he compared heights. Instead, he compiled lists
of extraordinary people and found that they were often related, or, in
Galton's memorable phrase, "Characteristics cling to families."

His 1869 book, *Hereditary Genius: An Inquiry Into Its Laws and
Consequences*, begins, "I propose to show in this book that a man's
natural abilities are derived by inheritance." Galton supported his
conclusion by documenting many cases of extremely accomplished
relatives. He began with 850 Englishmen over the age of 50 who were
listed in the biographical book, *Men of the Time*. He then culled this
list to 500 men who were especially well known. He estimated that
there were two million British men over the age of 50, so these 500
were about one in 4,000.

Their diverse talents were reflected in Galton's eclectic chapters:
judges, statesmen, commanders, literary men, men of science, poets,
musicians, painters, diviners, Senior Classics of Cambridge, oarsmen,

and wrestlers of the North Country. Galton meticulously searched for relatives of these 500 eminences, combing lists of accomplished people and obituaries. The results were reported in appendixes for each chapter that showed the eminent relatives, along with a brief biography of each.

Galton also looked at other countries and time periods. Johann Sebastian Bach was related to nineteen eminent musicians. John Milton's father was an accomplished musician and his brother was a judge. Charles Darwin's grandfather, Erasmus ("physician, physiologist, and poet") was considered to be even more eminent than Darwin. Galton might have added his own name, since he was a grandson of Erasmus. Instead, he listed several other relatives and modestly wrote that, "I could add the names of others of the family, who in a lesser but yet decided degree, have shown a taste for subjects of natural history."

Galton also argued that geniuses were being weeded out of the population—an apparent survival of the dummies. Galton reckoned that during the 100-year period from 530 B.C. to 430 B.C., 45,000 free men over the age of 50 lived in the city of Attica in Ancient Greece. In comparison to the size of the modern British population, there would be at most one super intelligent man in Attica; instead, there were four: Pericles, Socrates, Plato, and Phidias (not to mention Aristotle, who lived between 384 B.C. and 322 B.C., and was arguably the smartest of them all).

Galton speculated that too many supremely intelligent British men did not have children because of church requirements of celibacy, university fellowships conditioned on not marrying, and the imprisonment or death of those who were intelligent and brave enough to challenge the church or state. These theories are plausible, but another explanation is luck. Out of all civilizations in all historical time periods, some are bound to have several geniuses. By comparison, other civilizations in other periods will seem gripped by the mediocrity magnet—they regress to the mean.

Hereditary Intelligence

The first IQ (Intelligence Quotient) test consisted of 54 mental "stunts" that two psychologists, Alfred Binet and Theodore Simon, devised in 1904 to weed dull students out of the Paris school system.

Their intention was to gauge a person's general intelligence, including an accurate memory and the ability to reason clearly and logically—traits that would be required to succeed in school.

There are many different dimensions of intelligence, including the ability to remember, learn, reason, and solve problems. The Board of Scientific Affairs of the American Psychological Association observed that,

> Individuals differ from one another in their ability to understand complex ideas, to adapt effectively to the environment, to learn from experience, to engage in various forms of reasoning, to overcome obstacles by taking thought. Although these individual differences can be substantial, they are never entirely consistent: a given person's intellectual performance will vary on different occasions, in different domains, as judged by different criteria.

Some people argue that a measure of intellectual abilities should include musical intelligence, physical intelligence, and social intelligence.

We do not need to choose a specific definition of intelligence and a particular test designed to measure it. Whatever the definition and whatever the test, we can define a person's *ability* as the hypothetical average score if he or she were to take comparable tests innumerable times. No one will get the same score on repeated tests. Sometimes, a score will be higher than a person's ability (good luck); sometimes, it will be lower (bad luck). Therefore, individual scores regress to the mean, in that a person who gets a score that is exceptionally far from the mean will usually get a score closer to the mean if retested.

Someone who gets a 130 on one test will probably get a somewhat lower score on a second test. Someone who gets a 70 will probably get a somewhat higher score. It would be a mistake to conclude that the first person must be getting senile and the second person must be taking smart pills. It is more likely that their scores were fluctuating around their mental ability.

There is also regression when we compare the IQ scores of parents and their children. Very intelligent parents usually have pretty smart children, but there is not a perfect relationship. Look at any family with several children. The children may have many similarities, but they are not equally tall, equally strong, or equally intelligent. Therefore, we expect to see regression to the mean when we compare the IQ scores of parents and their children—just as there is regression to the mean when comparing the heights of parents and their children.

There is a positive correlation between the IQs of parents and their children because of genetics and because some environmental influences may be correlated with IQ; for example, high-IQ parents may provide a more nurturing environment that nourishes their children's IQ. That's fine. There will still be regression as long as the correlation between the IQs of parents and their children is not perfect.

As with heights, those with the highest IQs tend to have not only positive genetic influences, but also positive luck. Parents and their children may share genes, but they don't share luck. People whose IQs are far from the mean are likely to have had considerable luck; their parents and children are unlikely to have the same degree of luck. Therefore, IQ scores regress to the mean, whether we look forward or backward. Extreme parents tend to have less extreme children, and extreme children tend to have less extreme parents.

Am I just making this stuff up? Well, it turns out that there is a wonderful institute called the Henry A. Murray Research Center, named after a legendary Harvard psychologist, that has an extensive collection of data collected by Murray and other distinguished psychologists. Libraries have the papers these luminaries wrote. The Murray Center has the original data used in these papers. Best of all, the data are free for the asking, for anyone to reanalyze as they see fit. I contacted the Center and, sure enough, they gave me a collection of IQ test-scores for 43 married couples and their children, who were tested between the ages of three and ten years old.

Following Galton's study of the heights of parents and their children, I calculated the mid-parent IQ by averaging the mother and father's IQ. Figure 1 shows a scatter plot of these mid-parent IQs and their children's IQs. Points that are above the 45-degree line are cases where the child's IQ is higher than the parents' IQ; points below the line are cases where the child's IQ is lower than the parents' IQ.

Figure 1
Parents with Abnormal IQs Tend to Have Children with IQs Closer to the Mean

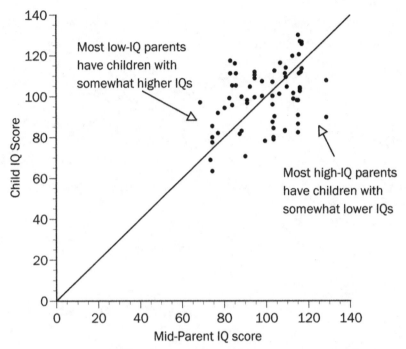

It should be no surprise that parents whose IQs are far from the mean tend to have children whose IQs are closer to the mean. The ten parents with the highest mid-parent IQ had an average IQ of 118, while their children had an average IQ of 111. The ten parents with the lowest mid-parent IQ averaged 76, while their children averaged 84.

Just as abnormal parents tend to have less abnormal children, so abnormal children tend to have less abnormal parents. Figure 2 confirms this by flipping the axes in Figure 1. Now, points above the 45-degree line are cases where the parents' IQ is higher than the child's; points below the line are cases where the parents' IQ is lower than the child's. Children with exceptionally high or low IQs tend to have parents who are closer to average. Specifically, the ten children with the highest IQ averaged 121, while their parents had an average IQ of 108. The ten children with the lowest IQ averaged 76, while their parents averaged 86.

Figure 2
Children with Abnormal IQs Tend to Have Parents with IQs Closer to the Mean

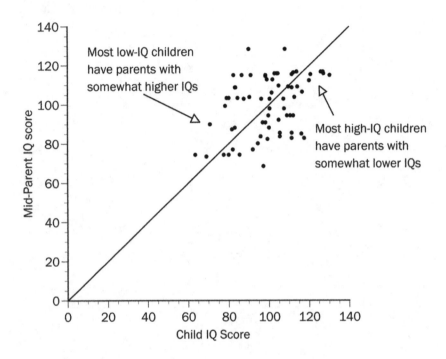

It is tempting to concoct a causal explanation for the fact that extremely intelligent people usually have children who are less extraordinary. Perhaps geniuses are bad parents and this depresses their children's intelligence. Perhaps the children of accomplished people do not develop their own intellectual abilities because they fear that they will look bad in comparison to their parents. Maybe, but there is also the purely statistical explanation of regression to the mean.

Bright Women and Duller Husbands

One of the most impressive students I ever had the privilege to teach was a young woman I'll call Ivy. She excelled in her coursework and we even co-authored a peer-reviewed paper while she was still an undergraduate. Ivy won a prestigious scholarship for postgraduate study in England and then went to Stanford Law School where she served on the Law Review. After Stanford, Ivy clerked for a Supreme Court Justice. She is now herself a law professor.

Ivy married a nice guy, but he is arguably less intelligent and certainly less accomplished. This is not an aberration. I made a list of the half-dozen most intelligent female students I've known who are now married. It is not a perfect or exhaustive list, but these women are all spectacularly talented. Five married men less intelligent than themselves; for the sixth, it could go either way. Their husbands are not dummies, but they are not as impressive as their wives. On a scale of 1 to 10, the women are 10s and their husbands are 7 to 9.

Why is this? Are all-star women threatened by intelligent men? Do these women want the upper hand in spousal debates? Do they need to feel superior at home as well as at work? The most likely explanation is, once again, regression to the mean.

There is a positive correlation because intelligent people typically prefer the company of people of comparable intelligence, but the correlation is not perfect because other things matter, too. Intelligent women may be attracted to intelligent men, but they also may seek partners who are caring, funny, sexy, whatever. There is an imperfect correlation between spousal intelligence and, so, women whose intelligence is far from the mean tend to have spouses who are somewhat less intelligent. It works in the other direction, too. Men whose intelligence is far from the mean tend to have spouses whose intelligence is closer to the mean.

The Murray Research Center data confirm this. The top quarter of the female IQs averaged 119 while their husbands averaged 109. Looked at the other way around, the top-quartile men averaged 117 while their wives averaged 107.

Athletic Abilities

Every town has athletic families—physically gifted children, often nourished and coached by parents who were quite the athlete in their day. We've all seen the coach whose child is the team star. (You've probably also seen the dark side of "daddy ball"—parents who coach to ensure that their not-so-talented children are treated like stars.)

At the national level, there have been many memorable athletic dynasties. In the early years of baseball, the five Delahanty brothers (Ed, Frank, Jim, Joe, and Tom) all played in the major leagues. Ed was elected to the Hall of Fame and has the fifth highest career batting average in major league history. In the modern era, the three DiMaggio brothers (Joe, Vince, and Dominic) played major league baseball, as did the Molina brothers (Bengie, Jose, and Yadier), and the Alou brothers (Felipe, Matty, and Jesus). Oddly (or, perhaps, not so oddly), all three DiMaggio brothers played center field. The Alou brothers had an even more unlikely coincidence. They all played for the San Francisco Giants in 1963 and, in one game, all three were in the outfield at the same time—Matty in left, Felipe in center, and Jesus in right. In another oddity that has never happened before or since, the three Alous all batted in one inning (unfortunately, Jesus grounded out, Matty struck out, and Felipe grounded out).

There have also been remarkable parent-child combinations. Chapter 1 noted that Peyton Manning, one of the greatest football quarterbacks ever, has a brother, Eli, who is also a pretty good professional quarterback, and their dad, Archie Manning, was a terrific quarterback even though he played on dreadful teams throughout his professional career.

Ken Griffey, Sr., was three times a Major League Baseball All-Star; his son, Ken Griffey, Jr., was thirteen times. In 1990, when Ken, Sr. was 40 and Ken, Jr. was 20, they hit back-to-back home runs while playing for the Seattle Mariners. Bobby ("The Golden Jet") Hull was a National Hockey League All-Star twelve times. His son, Brett ("The Golden Brett") was an All-Star nine times. Both are in the Hall of Fame. Ken Norton, Sr., was a heavyweight boxing champion and once broke Muhammad Ali's jaw. Ken Norton, Jr., was a hard-hitting linebacker who played on three Super Bowl championship teams and was selected for the Pro Bowl three times. Calvin Hill was a four-time All-Star in the National Football League. His son, Grant Hill, was a seven-time All-Star in the National Basketball Association.

These stories are memorable because they are exceptions to rule. If the children of great athletes were invariably great athletes, we wouldn't think anything of the Mannings, Griffeys, and Hulls. Ironically, because it is reasonable that athletic parents have athletic children, and because we know several families where this is true, there is a temptation to think that it is always true. It isn't. There have been thousands of extraordinary athletes who children are less extraordinary.

There is a general human tendency to have selective recall. We remember things that support our beliefs and discount evidence to the contrary. I tried to minimize my own selective recall by making a list of all the high school children in my town who either are exceptional athletes or have parents who were exceptional athletes. There is a rough positive correlation. The athletic children generally have athletic parents and the athletic parents typically have athletic children. But there is also the mediocrity magnet. The most athletic children are more athletic than their parents and the most athletic parents are more athletic than their children.

One dad played professional baseball. His son is good, but almost certainly will not go pro. Another dad set scoring records in high school and college basketball. His son is above average at best. The boy who is the best basketball player has a father who was pretty good, but not great. I have a son who is one of the top athletes in multiple sports. The best I can say about his father is that, for a college professor, he was a pretty good athlete—which is like being the world's tallest midget.

There is a double regression-to-the-mean whammy. First, just as highly intelligent people tend to marry people less intelligent than themselves, so exceptionally athletic people tend to marry people less athletic than themselves. Mid-parent athleticism regresses to the mean. Second, the athleticism of the children of parents whose mid-parent athleticism is far from the mean also regresses to the mean.

One might think that putting together two extraordinary athletes (like two Olympic gold medalists) would produce children who are even more extraordinarily gifted than their parents, sort of like combining a powerful engine with an aerodynamic design to produce a fast car. It doesn't work that way with athletic prowess, because of regression. The children of exceptional athletes will typically be above average, but not as far above average as their parents, and they seldom are more exceptional than their parents.

III. EDUCATION

4
Testing 1, 2, 3

SUPPOSE THAT I WERE TO TEST YOUR KNOWLEDGE OF WORLD HISTORY by asking you 20 true-false questions; for example:

Alexander Hamilton was a U.S. President, true or false?

What would be your test score? Relax, I'm not going to test you. But imagine that I did. We can think of your score—or anyone's score—as having two components—ability and luck. Your "ability" is your hypothetical average score if you were given a very large number of comparable tests.

Your ability is 80 if you would average 80 percent correct, but you won't get 80 on every single test. There is luck involved in how you feel when taking the test ("I think I'm coming down with something"), whether your guesses turn out to be correct ("Alexander Hamilton might have been president"), and, most fundamentally, in the questions that are asked ("I don't know much about early U.S. history").

Imagine a test bank with zillions of questions, from which 20 questions are selected for a test. By the luck of the draw, a person with an ability of 80 may score 90 on one test, but only 70 on another test. The score on any single test is an imperfect measure of ability. It's not just history tests. Luck is endemic whenever a test is used to measure a person's aptitude.

We observe scores, not ability. If luck is involved, what, if anything, can we infer from test scores? A key insight is that someone whose test score is high relative to the other people who took the test probably had good luck in that the score is probably higher than the person's ability. Someone who scores 90 could be someone of more modest ability (perhaps 85, 80, or 75) who did unusually well or could be someone of higher ability (perhaps 95) who did poorly. The former is more likely because there are more people with ability below 90 than above it.

If this person's ability is, in fact, below 90, then when this person is retested, his or her score will probably be below 90. Similarly, a person who scores far below average is likely to have had an off day and should anticipate doing somewhat better on later tests. This tendency of people who score far from the mean on one test to score closer to the mean on a second test is regression toward the mean.

To make this argument more concrete, consider the hypothetical abilities and scores in Figure 1 for 45 people who take a standardized test that has a maximum score of 800. These 45 people have abilities ranging from 550 to 750, with an average of 650. I assumed that scores

Figure 1
Scores Depend on Abilities

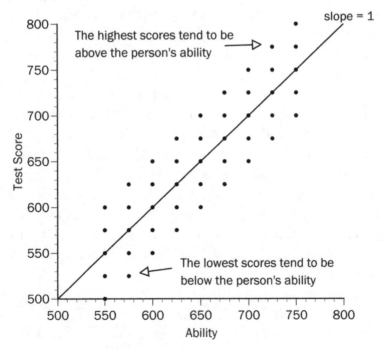

are evenly distributed about abilities, with an average luck of zero. The uniform distribution of abilities is unrealistic, as are the severely limited values. This stark simplicity is intended to clarify the argument.

The best-fit line in Figure 1 goes through the origin and has a slope of one, which means that ability is the best predictor of a person's test score. There is, however, an interesting pattern to the scores. Scores above the line are higher than ability (positive luck), while scores below the line are lower than ability (negative luck). Notice that the people with the highest scores tended to have had positive luck, and scored above their ability, while the people with the lowest scores tended to have had negative luck and scored below their ability.

Specifically, the five people with the highest test scores averaged 767, but their average ability is only 733, and they probably won't do as well on another test of the same material. It doesn't matter whether the second test is given before or after this test. All that matters is that they probably won't be as lucky on the second test as they were on this test.

Figure 2
Estimating Abilities from Scores

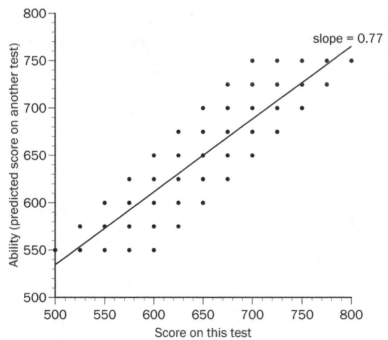

On the low side, the people with the lowest scores have somewhat higher ability and probably won't do as poorly on a second test because they won't have as much bad luck as they had on this test.

Figure 2 shows this regression graphically by reversing the axes so that scores are now used to assess ability. Because ability is the best predictor of the score on another test, we can consider the vertical axis to be both a person's ability and the predicted score on another test.

The best-fit line has a slope of 0.77, which is less than 1 because of regression. Someone who scores 100 points above average is predicted to score only 77 points above average on another test on the same subject with the same difficulty. The same is true of below-average scores. Someone who scores 100 points below average is predicted to score only 77 points below average on a second test.

The larger the role of luck relative to ability, the larger the regression to the mean. Figure 3 shows the relationship between scores and abilities when the luck factor is doubled. The best-fit line flattens and the slope falls to 0.45. Now, a person with a score of 100 points above or below average is predicted to score 45 points above or below average on a second test.

Figure 3
Predicting Abilities from Scores That Are More Influenced by Luck

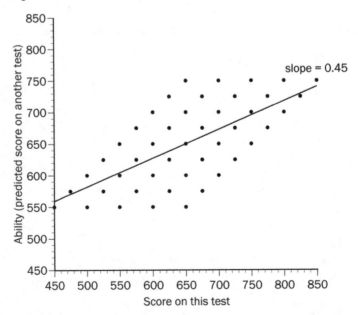

Figure 4
Predicting Abilities from Scores that Are All Luck

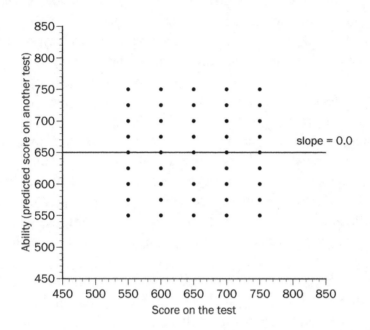

The line is even flatter for tests in which ability matters less and luck more. The extreme is where the scores are all luck. Suppose that the test is machine-graded and the machine malfunctions. The reported test scores are just random numbers with no relationship at all to a person's ability. Figure 4 shows that if the scores are all luck, the regression line is horizontal because the test scores do not help us assess ability. Not being able to distinguish one person's ability from another's, our best guess of anyone's ability is the average score, 650.

At the other extreme is the mythical perfect test with no luck, so that a person's score is the same every time, test after test, and there is no regression to the mean. In the real world, tests are imperfect and test scores regress.

Some Real Test Scores

Regression in test scores isn't just theoretical. The students who score highest on the midterm test in my classes usually do not do quite as well on the final examination, while those who score lowest on the midterm generally do somewhat better on the final. Are my students converging to a depressing mediocrity as the semester progresses, with the weak students learning and the strong students forgetting? Maybe the low scorers are energized by a fear of failing the course and the high scorers become complacent. That's certainly possible. However, it is also true that the students with the highest scores on the final usually got somewhat lower scores on the midterm, while those with the lowest scores on the final did somewhat better on the midterm. Since the final exam occurs after the midterm, it is hard to see how a stellar or dismal performance on the final could have affected their preparation for the midterm.

A plausible explanation is regression to the mean, which works in either direction because it predicts that those who obtain the highest (or lowest) scores on either test will be more nearly average on the other test, regardless of which test is taken first.

Those students with the highest scores on either test are mostly above-average students who did unusually well because the questions asked happened to be ones that they were well prepared to answer. They are generally good students who did unusually well, not great students who had an off day. On average, most won't score so high on another test.

Figure 5 shows a scatter plot of the midterm and final exam test scores in an introductory statistics class that I taught twelve times in a ten-year period. If there were no regression, the scores would be evenly scattered about the 45-degree line drawn in the figure. There is, in fact, a pattern. Scores above the 45-degree line are students who did better on the final than on the midterm; scores below the line are students who worse on the final than on the midterm. The students with the highest scores on the midterm are overwhelming below the line (they did not do as well on the final), while the students with the lowest scores on the midterm are generally above the line (they did better on the final).

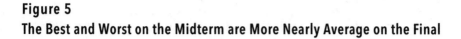

Figure 5
The Best and Worst on the Midterm are More Nearly Average on the Final

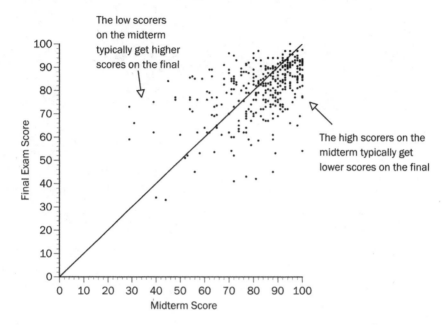

Specifically, the 10 percent of the students with the highest midterm scores averaged 99 on the midterm and 87 on the final; the 10 percent with the lowest midterm scores averaged 54 on the midterm and 69 on the final.

Figure 6 reverses the axes. Now, the high scorers on the final tend to be below the line (they did not do as well on the midterm), while the low scorers on the final are generally above the line (they did better on the midterm). Here, the 10 percent of the students with the highest final exam scores averaged 95 on the final and 92 on the midterm; the 10 percent with the lowest midterm scores averaged 55 on the final and 68 on the midterm. Test scores do regress toward the mean.

The observed improvement in the scores of those who did poorly on the midterm and the decline in the scores of those who did well may not be merely statistical. The low-scorers may study more and the high-scorers may coast. The next chapter will explain how we can separate out the regression effect and determine if abilities are changing from one test to the next.

Figure 6
The Best and Worst on the Final are More Nearly Average on the Midterm

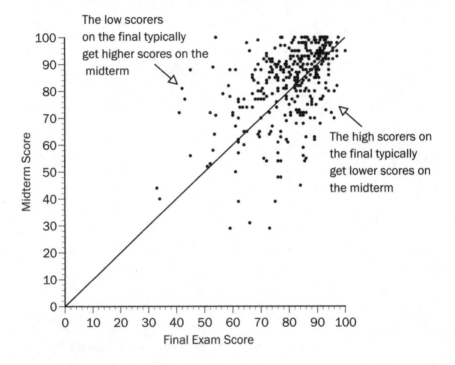

Asymmetric Information

In my statistics classes, 20 percent of the course grade is based on the midterm and 40 percent on the final exam. After the midterm (but before the final), I offer every student the option of throwing out the midterm score and counting the final exam as both the midterm and final (in effect, 60 percent of the course grade). My reasoning is that, yes, test scores are an imperfect measure of ability and, yes, students know more about their abilities than I do. Those students who believe that they had bad luck on the midterm can bet their grade that they are right. Those who believe their midterm scores are above their abilities can hold on to their scores and be thankful that they were lucky.

This offer is a practical example of asymmetric information, when two people agree to something but one person knows more than the other about the situation. For example, an insurance company might sell life insurance to Justin, who the company assumes is of average health, while Justin knows his health is fragile. My midterm/final exam offer is more benign and has a built-in incentive to study harder.

Students with grades of C or lower generally take the offer, though I once had a hopelessly misguided student who asked me if he should hold onto his failing midterm grade or gamble on the final. Occasionally, students with midterm grades of B, B+, or even A- will throw out the midterm because they are convinced that these scores understate their abilities.

In practice, about 60 percent of the students who take the gamble win their bets, doing better on the final exam than they did on the midterm. The other 40 percent evidently have a misplaced belief in their abilities in that they do worse on the final than they did on the midterm.

I like my offer because there are consequences if a student discards the midterm and then does poorly on the final. No one is allowed a flip-flop after the final exam: "Never mind, I decided to keep my midterm score. Oh, and by the way, can I throw out my final exam?" Not so with the Scholastic Aptitude Tests (SATs). Students can retake the SATs as many times as they want and many colleges only consider the student's highest scores, which are a biased measure of the student's aptitude. It would make more sense to look at a student's average score.

5

The Beatings Will Continue Until Morale Improves

KAHNEMAN AND TVERSKY IDENTIFIED A HUMAN INCLINATION TO rely on a reference point when making decisions, a tendency they called "anchoring." A student term paper in one of my statistics classes illustrated this foible. Randomly selected students were asked one of these two questions:

> The population of Bolivia is 5 million. Estimate the population of Bulgaria.

> The population of Bolivia is 15 million. Estimate the population of Bulgaria.

Those who were told that Bolivia's population was 15 million tended to give higher answers than did those who were told that Bolivia's population was 5 million. Several similar questions confirmed this pattern. People use a known "fact" as an anchor for their guesses.

Car dealers use anchoring to manipulate us into paying high prices. We are inclined to judge whether we are getting a good deal by comparing the final negotiated price to the dealer's initial price, no matter how unrealistic the initial price. Thus, an experienced car salesman starts the haggling with a high price.

Another human foible is an ignorance of regression to the mean. Kahneman once tried to convince Israeli flight instructors that trainees would progress faster if they were praised instead of punished. A senior instructor objected, telling Kahneman that,

On many occasions I have praised flight cadets for clean execution of some aerobatic maneuver, and in general, when they try it again, they do worse. On the other hand, I have often screamed at cadets for bad execution, and in general, they do better the next time. So please don't tell us that reinforcement works and punishment does not, because the opposite is the case.

Perhaps the pilots who were praised for doing well became complacent and less attentive, while those who were screamed at for doing poorly became more focused and energized by a fear of being kicked out of the training program?

Kahneman knew that there was another, completely different explanation. He drew a target on the floor with chalk and asked each instructor to turn his back to the target and throw two coins, one after the other, at the target. Kahneman measured the distance of each coin from the target. Some officers did better than others, but the officers who were closest on their first toss tended to do not as well on their second toss, while the reverse was true of those who were farthest away on their first toss. Did their ability somehow change between the two tosses, with the best tossers becoming complacent and the worst tossers more focused? Not likely, because none of the officers looked at the result of the first toss before making the second toss.

The clincher is that if we analyze the data in reverse order, those who did best on their second toss generally did worse on their first toss, while those who did poorly on their second toss usually did better on their first toss. How could the second toss possibly have influenced the first toss?

The answer to this seeming paradox is that those officers who did well on either toss generally had some good luck in the angle and force with which they tossed the coin and the way the coin bounced and rolled after landing. Unusual luck cannot be counted on to repeat, so they typically did not do as well on their other toss. Similarly, those officers who did poorly on one toss were generally unlucky on that toss and not so unlucky on the other toss.

In the same way, some pilots surely have more ability than others, but there is also an element of luck involved in that, no matter what their ability, each pilot has flights that are better or worse than others. Those cadets who had the best flights compared to other pilots are more likely to have had good luck than bad luck, and are generally not as far above average in ability as they were in performance. They will usually not do as well on their next flight, whether the instructor praises them, screams at them, or says nothing at all. The senior instructor who chastised Kahneman mistakenly thought that his praise made the cadets do worse when the truth is that they were not as good as they seemed to be. Similarly, the cadets who had the weakest flights were, on average, not as incompetent as they seemed and should do better on their next flight even if the instructor can control his screaming.

Kahneman later related that, "I knew that this demonstration would not undo the effects of lifelong exposure to a perverse contingency." Nonetheless, the experience was eye-opening:

> This was a joyous moment, in which I understood an important truth about the world: because we tend to reward others when they do well and punish them when they do badly, and because there is regression to the mean, it is part of the human condition that we are statistically punished for rewarding others and rewarded for punishing them.

Pilot Training Scores

Reid Dorsey-Palmateer (a former student and now a professor) and I investigated Kahneman's memorable anecdote with actual U.S. Navy flight training data. These data demonstrate how flight scores are susceptible to regression and, more importantly, show how to assess changes in a pilot's ability as the training proceeds.

Our data came from the final stage of a six-phase aviation training track for naval aviators. In this carrier qualifying phase, the pilots practice solo landings on a Naval aircraft carrier under the supervision of an officer stationed on the carrier who grades each flight and debriefs the pilot afterwards.

In a carrier landing the pilot aims for the middle one of three arresting wires on the deck of an aircraft carrier. Here is Lieutenant Corey Johnston's description:

> The day time landing pattern begins overhead the ship at anywhere from 2000-5000 feet. When aircraft are able to land, they circle and descend to establish themselves at 800 feet and three miles behind the aircraft carrier at a minimum of 250 knots. As the aircraft passes over the carrier, they will go into a level 2G break turn to the left to bleed energy and get to the airspeed required to drop flaps and landing gear. The aircraft will drop landing gear, flaps, and arresting hook and descend to 600 feet on a down wind track approximately one mile abeam the ship. When the aircraft is abeam the aft portion of the ship, they start a descending 180 degree left hand turn. The turn is conducted at 20-22 degrees angle of bank, starting at 200-300 foot-per-minute rate of descent. After 90 degree of turn, the aircraft will be at 450 feet and increase the rate of descent to 500-600 feet-per-minute. The pilot will maneuver the aircraft to be at 325-375 feet crossing the ship's wake and begin to visually assess where he is on glideslope. This will take him to the start of the pass. The pass begins when the aircraft rolls out on centerline of the carrier. . . .
>
> In civilian and air force flying, the aircraft is flared to slow down and land. In carrier flying, the aircraft is kept at a constant altitude all the way around the turn to landing. If the aircraft nose is pitched down, we call the aircraft fast. If the nose is pitched up, we call the aircraft slow. A fast aircraft will raise the arresting hook up possibly causing the aircraft to miss all of the wires. We call this a bolter. If the aircraft does engage the wires, there is a possibility of overstressing the arresting gear motors or breaking the arresting hook off of the aircraft. If the aircraft gets slow, there is a possibility of stalling the aircraft (running out of lift over the wings) and crashing.

Carrier landings are clearly complicated and challenging. Any pilot, no matter how talented, will not perform equally well on every landing.

Table 1 shows the scores for 1,828 graded flights for ten different pilots tested over a two-week period. There were no scores of 0 (unsafe), 2.5 (bolter), or 5 (perfect). The average score was 2.71.

Table 1
Landing Scores

Score	Meaning	Percent of Landings
5.0	perfect pass	0.0
4.0	reasonable deviations with good correction	6.2
3.0	reasonable deviation	57.7
2.5	bolter	0.0
2.0	below average but safe pass	30.6
1.0	waveoff	5.6
0.0	unsafe, gross deviation	0.0

We calculated each pilot's average score for the day and compared these averages across successive days. One of our research questions was whether scores regress to the mean. Another was how we might use scores to assess changes in abilities from one day to the next.

Table 2 shows the movements between flight scores. It was not literally true that every high score was followed by a lower score and every low score was followed by a higher score. In fact, 48 percent of the time, the scores on consecutive landings were identical—no doubt a reflection of the fact that even though there is a continuum of performances, only four scores were given (1, 2, 3, or 4) and 88 percent of these scores were 2s and 3s. Still, as Kahneman observed, there is clearly regression. For the scores at the extreme, there is nowhere to go but toward the middle and most did: 77 percent of the 4s worsened and 83 percent of the 1s improved. Of the 3s, 32 percent did worse and 7 percent did better on the next landing. Of the 2s, 58 percent did better and 7 percent did worse.

Table 2
Movements Between Scores

Previous Score	Current Score			
	1	2	3	4
1	17	40	44	0
2	35	187	283	18
3	42	279	602	65
4	1	17	58	23

Estimating Ability From Performance

Truman Kelley was a professor at the Harvard School of Education when he wrote a 772-page textbook in 1947 entitled *Fundamentals of Statistics*. Buried deep in this dense book is a remarkable formula that has become known as Kelley's equation. His formula says that the optimal prediction of a person's ability is a weighted average of the person's performance and the average performance of the group the person belongs to:

estimated ability = R(performance) + $(1 - R)$(average group performance)

The term R is reliability, which measures the extent to which performances are consistent. If a group of students take two comparable tests, reliability is the correlation between their scores on these two tests.

Suppose that a test is administered to all high school seniors and we know nothing about the individual test takers. A student's performance is his or her score. The average group performance is the average test score. So, Kelley's equation is:

estimated ability = R(score) + $(1 - R)$(average score)

If the test scores were completely random, like guessing the answers to questions written in a language the students don't understand, the reliability would be zero, and our best estimate of a person's ability would be the average score of the group.

At the other extreme, a perfectly reliable test would be one where some students do better than others, but any single student gets the same score, test after test. Now, the best estimate of a student's ability is the student's score. This, too, makes sense.

In practice, tests are not completely worthless or 100-percent reliable. I just used these silly extremes to demonstrate that Kelley's equation makes sense for the lowest and highest possible values of reliability. If it didn't make sense for these values, it would be a flawed rule.

If the correlation between scores on comparable tests is 0.80, the reliability is 80 percent. In this case, Kelley's equation says that a person's estimated ability is 80 percent based on the person's score and 20 percent based on the average score:

estimated ability = (0.80)(score) + (0.20)(average score)

Suppose the average score is 60. A person who gets the average score is estimated to have average ability. Those who get above-average scores are estimated to have above-average ability, but their estimated ability is closer to average than is their score. For example, a person who scores 90 has an estimated ability of 84. Similarly, those with below-average scores are estimated to have below-average ability, but not as far below average as their score. A person who scores 30 has an estimated ability of 36.

If this sounds like regression to the mean, you're right. A performance that is far from the mean probably reflects ability that is closer to the mean. The wonder of Kelley's equation is that it tells us how much closer!

Although Kelley discovered the equation named after him using the thoroughly conventional approach to statistics that prevailed in the 1940s (and for many decades afterward), Kelley's equation can also be derived using the Bayesian logic that has become very popular in recent years.

The Bayesian approach begins with a *prior* estimate of a person's ability, before the test is taken. The revised estimate, taking the test result into account, is called the *posterior* estimate, with the relative weights given to the test and the prior depending on the test's reliability,

posterior estimate of ability = R(score) + (1 − R)(prior estimate of ability)

If we have no reason to think that a student's ability is above-average or below-average, our prior estimate of the student's ability might be the average test score, which makes the Bayesian estimate of the student's ability equivalent to Kelley's equation:

posterior estimate of ability = R(score) + (1 - R)(average score)

The nice thing about a Bayesian interpretation is that it helps us understand why Kelley's equation works. Before we give the test, we have no way of knowing whether a student is better or worse than average. After the performance, we can judge whether a student is above or below average based on the test score, but we know from the regression-to-the-mean argument that exceptional performances (good or bad) typically come from people whose abilities are not as exceptional as the performance. So, we revise our assessment of the student's ability towards the performance, but not all the way to the performance. In addition, a Bayesian approach allows us to keep revising our estimate of a student's ability if the student takes additional tests. Let's see how this works.

The Application of Kelley's Equation to Pilot Training Scores

I will use the Navy flight training data to show how Kelley's equation can be used to estimate ability from scores, and how our ability estimates can be revised as more information is acquired from additional scores.

The correlation between scores on consecutive days is 0.51, so this is our estimate of the test's reliability. Pilot 10's score was 2.78 on the first training day, when the average score was 2.35, so his estimated ability is 2.57:

$$\text{estimated ability} = R(\text{score}) + (1 - R)(\text{average score})$$
$$= 0.51(2.78) + (1 - 0.51)(2.35)$$
$$= 2.57$$

Pilot 10 scored well above average (2.78 versus 2.35) but, taking regression to the mean into account, we estimate his ability to be closer to average.

Going into the second day, our prior estimate of Pilot 10's ability is now 2.57. If he scores above 2.57, we revise our estimate of his ability upward. If he scores below 2.57, we revise our estimate downward.

As it turned out, Pilot 10's second day score was 2.71, slightly lower than his first-day score of 2.78. An outraged flight instructor might conclude that this pilot did worse because he was praised, but the 2.71 score is actually good news. We predicted that his score would fall from 2.78 to 2.57. The fact that he scored 2.71 means that we should revise our estimate of his ability upward. Instead of being disappointed that his score fell slightly, we should be encouraged by the fact that his score did not fall as much as expected.

Kelley's equation tells us that, based on Pilot 10's 2.71 score on the second day, we should revise our estimate of Pilot 10's ability upward from 2.57 to 2.64. Each day, Pilot 10's ability estimate is revised upward or downward depending on whether his score that day is above or below the most-recent ability estimate.

Figure 1
Scores and Ability for Pilot 10

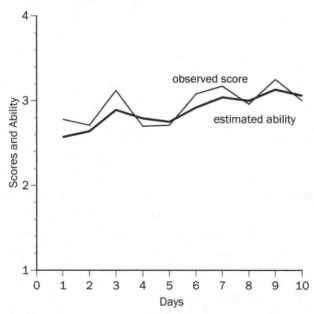

Figure 1 shows Pilot 10's scores and estimated ability through ten days of testing. By the tenth day, this pilot's estimated ability had risen from 2.57 to 3.06. Nine of the ten pilots showed an increase in ability over the course of the training period. The exception is Pilot

9, shown in Figure 2, whose ability was initially estimated to be 2.45, was revised upward for a while, and then was revised downward to 2.45, the same as on the first day of training.

Figure 2
Scores and Estimated Ability for Pilot 9

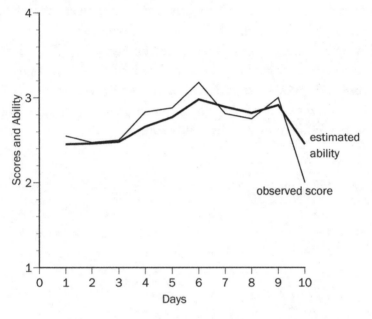

This is a great example because it confirms Kahneman's anecdote and it also shows how ability can be estimated while taking regression into account. If those pilots who are initially above-average slip less and those who are initially below-average improve more than predicted, the training is succeeding. In this particular case, nine of the ten pilots improved and the tenth showed no evidence of a change in ability.

Kelley's equation is extraordinarily useful because it not only recognizes that performances regress to the mean, but it tells us how much regression to expect. Kelly's equation shows us how to estimate ability from performance, and how to revise our ability estimates based on additional performance data.

It works for pilot training. It works for educational testing. It works for sports and business. It works whenever luck—good or bad—causes performances to be an imperfect measure of ability.

6

Money For Nothing

WAS CONTACTED BY A BOARD MEMBER ("EMMA") FOR AN EXPENSIVE private school ("GreenLeaf") which offers classes from kindergarten through eighth grade. Emma told me that GreenLeaf was located in an affluent suburb, with very good public schools. Nonetheless, GreenLeaf's headmaster was able to convince parents that they should spend more than $10,000 annually to obtain a GreenLeaf education.

Emma said that two kinds of parents send their children to GreenLeaf. One is rich families who like to network with other rich families and want a safe environment where their children won't be bullied or led astray by "bad kids" who go to public schools. The second target audience is immigrant families who believe that, as was true in the countries they left, children have to go to private schools to get a good education.

GreenLeaf's headmaster appealed to both types of parents with a slick PowerPoint presentation that showed its students engaged in wholesome activities and boasted that the students consistently score in the 90th percentile on national standardized tests.

Emma had gone to a family get-together and mentioned these high test scores to a relative ("Rachel") who had a PhD in education and was an expert on charter schools. Rachel said that nationwide comparisons can be misleading because nationwide scores are pulled down over time relative to suburban schools by the poor performance of many rural and big-city schools. Why was GreenLeaf using national scores, rather than suburban scores, when it was drawing students from suburban school districts?

Emma said, "Tell me more," and Rachel went on. Standardized tests have long been used to assess individual students; for example, to identify student deficiencies and to separate students into classes appropriate for their abilities. Today, however, many states use standardized tests to evaluate teachers and schools.

This isn't necessarily fair because students are not randomly assigned to schools. Parents choose to live in certain neighborhoods, send their children to certain schools, and enroll them in certain classes. These choices may depend on family income and the parents' interest in their children's education.

Average test scores are consequently a misleading measure of the effectiveness of teachers and schools. Some classes have higher test scores than other classes and some schools have higher test scores than other schools not because of the teacher or the school, but because of the students.

Recognizing this, many state criteria for evaluating teachers and schools look at changes in test scores over time. This, too, is problematic if the student population changes from year to year. This year's sixth graders may have higher average scores than last year's sixth graders because this year's sixth graders are better students than were last year's sixth graders. This year's sixth graders may have higher average scores than last year's fifth graders because weaker students left the school.

In addition, as we now know, test scores regress to the mean. If this statistical regression is not taken into account, changes in tests scores may be misinterpreted as changes in ability rather than fluctuations in scores about ability. All of these issues are likely to be especially acute for schools like GreenLeaf that have a small number of students, selective admissions requirements, and substantial student turnover.

The next day, Emma called me and asked me to do a statistical analysis of GreenLeaf's test scores. I did a little background work. Many of the parents I talked to said that they send their children to GreenLeaf because of the high test scores. In fact, the curriculum is reputed to be a grade ahead in that, for example, the school's fourth-grade students are doing work comparable to fifth-grade students in the local public schools. (Left unmentioned is that the kids are also an age ahead in that the school's fourth grade students are, on average, as old as the fifth-grade students in public schools.)

I also learned that twenty students (ten boys and ten girls) are admitted to the kindergarten class each year, with an assessment test an important component of the admissions process. After the students are admitted, Educational Records Bureau (ERB) tests are administered each year from first grade through eighth grade to assess each student's progress.

Over time, students leave the school for a variety of reasons, and are replaced by new students as the school tries to maintain a class of twenty at every grade level, evenly divided between males and females. To get a meaningful assessment of GreenLeaf's test scores, I realized that I would have to take into account the possibility that the students entering GreenLeaf might be stronger academically than the students they are replacing. I needed individual scores for all the students so that I could distinguish between new students and those students who stayed at GreenLeaf from first grade through eighth grade.

The headmaster refused to turn over the scores, arguing that they were confidential and, in addition, that some scores had been destroyed and others had been put into storage and could not be retrieved easily. Emma persuaded GreenLeaf's board to give her access to the stored scores, and she spent an afternoon sifting through boxes of them. She was able to find a complete set of scores for grades one through eight for the graduating classes of 2006, 2007, and 2008.

Emma protected the students' confidentially by identifying them with numbers instead of names. She was on a mission and wouldn't let go, especially because of the headmaster's suspicious refusal to turn over the scores.

Emma gave me the data and her suspicions were confirmed.

There is no way of knowing for certain how GreenLeaf students would have done if they had attended a competing school. There is a theoretical way of tackling this what-if question. Take a group of students who apply and are qualified for admission. Then flip a coin to determine which students are admitted and which aren't, and compare their progress over the next eight years. This wouldn't be popular, but it would be a controlled experiment.

As unlikely as it seems, this was actually done in the 1960s in Ypsilanti, Michigan, where children from low socioeconomic households were selected or rejected for an experimental preschool program based on the outcome of a coin flip. Without the coin flip, the study would have been tainted by the possibility that the parents who applied to the preschool program were systematically different from those who did not apply—perhaps more likely to be employed or more likely to care about their children's education. There is no way of knowing how the parents differed, so the next best thing is to separate the applicants randomly into a preschool group and non-preschool group, much like a medical study that randomly gives some people a medication and others a placebo.

The Ypsilanti study found that those who attended the preschool program were more likely to complete high school, less likely to be arrested, and more likely to have jobs. It may seem heartless to those who lost the coin flip, but this experiment demonstrated the value of the preschool program.

There had been no coin flips for admission to GreenLeaf, so I compared the performance of GreenLeaf students with the performance of students at competing schools.

I focused on the reading comprehension and mathematics scores, which are the only tests that are administered every year from first grade through eighth grade. Scores on the ERB tests are standardized using three populations:

National Norms: schools throughout the United States.

Independent School Norms: private schools in the United States and overseas.

Suburban School Norms: suburban public schools in the United States.

GreenLeaf is located in a suburban area with many excellent public schools. Good is not necessarily good enough to convince parents to pay thousands of dollars each year for a private-school education. They need to believe that GreenLeaf is substantially better than the local public schools. Thus, the suburban school norms are most appropriate for GreenLeaf.

My initial look at the data confirmed Rachel's warning about using national norms. Table 1 shows that in first grade, the first year GreenLeaf students take ERB tests, the GreenLeaf students are above the 90th percentile in comparison to either national or suburban schools. As the students progress through GreenLeaf, they stay at approximately the 90th percentile compared to national schools, but sink below the 70th percentile compared to suburban schools.

In first grade, a score that is in the 90th percentile for suburban schools is also in the 90th percentile for national schools. Students who are in the top ten percent at national schools would also be in the top 10 percent if they went to suburban schools. However, in later grades, scores that are in the 90th percentile for national schools are only in the 70th or 60th percentile for suburban schools. Students who are in the top ten percent at national schools would be only slightly above-average at suburban schools. This is an indictment of rural and urban schools, and it is also an indictment of GreenLeaf and the misleading statistics quoted by its headmaster.

Table 1
Greenleaf Student Percentiles Compared to National and Suburban Schools

| Grade | Reading Comprehension | | Mathematics | |
	National Schools	Suburban Schools	National Schools	Suburban Schools
1	94.5	91.2	98.4	90.2
2	87.1	68.0	94.9	77.2
3	87.4	67.9	90.9	73.2
4	90.3	73.0	91.3	77.7
5	87.3	66.1	92.4	72.8
6	83.3	62.9	89.7	76.8
7	80.0	62.2	91.0	68.8
8	88.0	62.5	92.8	64.2

There may be an innocent explanation for GreenLeaf's plummeting scores. Educational value added should be measured by *changes* in test scores rather than by the *level* of test scores, and such an assessment should take into account changes in the composition of the student body. We need to track the scores of those students who stayed at GreenLeaf from first grade through eighth grade (the "permanent" students) so that we are not misled by the replacement of weak students with stronger students.

In my data set, the number of students who completed all eight grades at GreenLeaf was nine in the class of 2006, thirteen in the class of 2007, and thirteen in the class of 2008. When I looked at the data, I found that the average scores were indeed artificially inflated by the continual replacement of weaker students with stronger students:

1.　　Students who dropped out of GreenLeaf generally had lower test scores than the students who stayed.

2.　　The students who replaced the departing students generally had higher test scores than the students who left *and* the students who stayed.

Table 2 shows that the decline in the scores of the students who stayed was even worse than suggested by the overall scores in Table 1. The drop in percentiles between first grade and eighth grade for the permanent students was from 93.0 to 54.0 in reading comprehension and from 94.5 to 63.5 in mathematics. Figure 1 shows these percentiles for students who were enrolled in all eight grades. The drop in scores is less pronounced in mathematics, but quite striking for both tests.

Table 2
Average Scores for All Students and for the Permanent Students

Grade	Reading Comprehension		Mathematics	
	All	Permanent	All	Permanent
1	91.2	93.0	90.2	94.5
2	68.0	69.0	77.2	81.1
3	67.9	70.8	73.2	80.2
4	73.0	70.4	77.7	83.9
5	66.1	60.9	72.5	71.8
6	62.9	58.4	76.8	78.4
7	62.2	55.8	68.8	68.1
8	62.5	54.0	64.2	63.5

Figure 1
Average Scores of Students Enrolled in All 8 Grades, Suburban School Norms

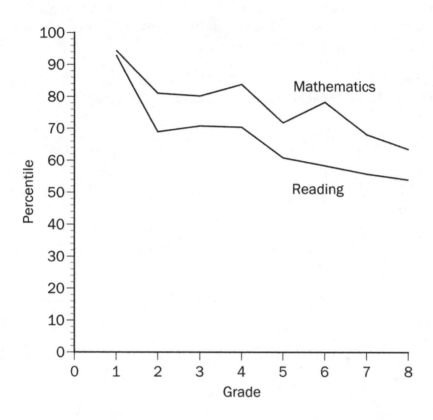

Perhaps regression to the mean is the explanation. GreenLeaf is a selective school that admits students based, in part, on their high test scores. We can consequently expect their abilities to be somewhat less exceptional than their test scores. If so, we expect these admitted students, on average, to score somewhat closer to the mean on subsequent tests. The question is, "How much closer?" If students who score in the 90th percentile on the admissions test are, on average, in the 80th percentile in ability, then we should not be surprised or disappointed if they score in the 80th percentile on a later test. We should be pleased if they score in the 90th percentile and disappointed if they score in the 70th percentile.

As with the naval pilots, we can use Kelley's equation to answer this question. The Educational Records Bureau reports 0.91 reliability values for its reading comprehension test and 0.85 for its mathematics test. Using these numbers, Kelley's equation can be used to estimate each student's ability, starting with the first-grade tests and continuing until the student graduates in eighth grade.

Figure 2
Estimated Average Ability of GreenLeaf Students Who Stay for All Eight Years

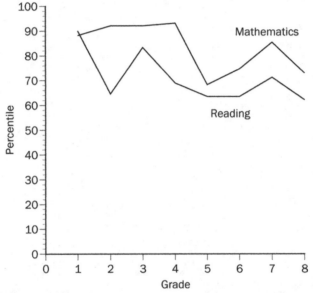

Figures 2 shows the average estimated ability of the students who stay at GreenLeaf, revised each year based on that year's test scores. In reading comprehension, the average estimated ability of the permanent students fell from the 90th percentile in first grade to the 62nd percentile in eighth grade. Thirty-four of 35 students had their estimated ability decline between first and eighth grade. In mathematics, the average estimated ability fell from the 88th percentile in first grade to the 73rd percentile in eighth grade. Thirty-two of 35 students had their estimated ability decline between first and eighth grade.

The results were what Emma feared. GreenLeaf parents may be getting something for all the money they are spending, but it is not a superior education.

7

Learning and Unlearning

MANY STATES REQUIRE SCHOOL CHILDREN TO TAKE HIGH-STAKES tests that are used by parents, school officials, and state administrators to assess students, teachers, and schools. In 1999, for example, Massachusetts public schools were given improvement goals for 2000 based on their 1999 scores on standardized tests. There was considerable consternation when most of the schools with the lowest scores in 1999 met their goals while many of the schools with highest scores did not; in fact, their scores often fell. It seemed that weak students were learning and strong students were unlearning.

However, part of the explanation for the improvements and setbacks was, no doubt, regression to the mean. I don't have detailed Massachusetts data, but I do have California data that illustrate the point.

California's STAR Program

We know that second graders who score in the 90th percentile are likely to be students of somewhat more modest ability who did unusually well, and that when these students take another test, they will probably not do as well. Students, parents, and teachers should anticipate this drop-off and not blame themselves if it occurs. Similarly, students who score far below average are likely to have had an off day and should anticipate doing somewhat better on later tests. Their subsequent higher scores may be a more accurate reflection of their ability rather than an improvement in their ability.

This reasoning applies to schools as well as students. Just as a single student who gets a very high score probably had good luck, it is also likely that a school that gets a high score probably had good luck. The questions asked on a particular test may be unusually well matched with the school's curriculum. Or perhaps the questions are ill suited for the neighborhood's culture. Or perhaps the test is given on a hot day and the school does not have air-conditioning. Or perhaps a school has been hit by an infectious disease and many students are feeling poorly. Or perhaps some students decide to stay home or not take the test seriously.

Let's see if the data support this reasoning. From 1998 through 2013 California's Standardized Testing and Reporting (STAR) program required all public school students in Grades two to eleven to be tested each year using statewide standardized tests. All schools were given an Academic Performance Index (API) score for ranking the school statewide and in comparison to one hundred schools with similar demographic characteristics. The API rankings were released to the media, displayed on the Internet, and reported to parents in a School Accountability Report Card. The scores are used for the state government's assessment of schools, the school administrators' assessment of teachers, and the parents' assessment of their children, teachers, and schools. The scores can have curricular effects on the schools and psychological effects on the families.

The API scores ranged from 200 to 1000, with an 800 target for every school. Any school with an API below 800 was given a one-year API growth target equal to 5 percent of the difference between its API and 800. Thus a school with an API of 600 had an API growth target of 610. The target for a school with an API above 800 was to maintain its API.

A school's API score was determined by the percentage of students in each of five quintiles established by nationwide scores for the tests. A truly average school that has 20 percent of its students in each quintile would have an API of 655, well below the state's 800 target. A Lake Woebegone school, with scores all above average and evenly distributed between the 50th and 99th percentile would have an API of 890.

(Garrison Keillor, host of the radio program *A Prairie Home Companion*, described the fictitious town of Lake Woebegone as a place where "all the children are above average." This is called the "Lake Woebegone Effect" by educational researchers to identify the flaw in claims that all schools should perform above average on state tests.)

Scores that have important consequences should be interpreted properly, taking into account any purely statistical reasons for changes in test scores over time. Yes, you guessed it. A comparison of school APIs across years shows regression toward the mean. Figure 1 shows that the top-performing schools in 2001-2002 tended to do not as well in 2002-2003, while the opposite was true of the bottom schools.

Figure 1
API scores in 2001-2002 and 2002-2003

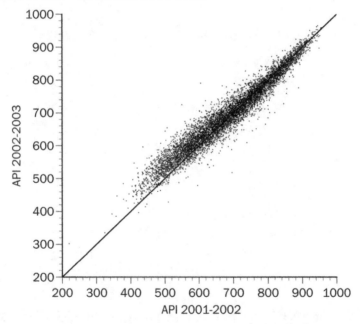

Overall, schools that were 100 points from the mean in 2001-2002 were, on average, only 87 points from the mean in 2002-2003. Kelley's equation predicts that a school with a 2001-2002 API of 550 will have a 2002-2003 API of 577. To put this into perspective, the growth target for a school with a 550 API is 563. Most low-performing schools can be expected to meet their targets without doing anything at all! On the other hand, a school with a 2001-2002 API of 850 is predicted to regress to 837 in 2002-2003 API, falsely suggesting that the school is doing a bad job.

The state forms groups of one hundred schools with similar characteristics and divides the schools in each group into deciles based on API scores. Table 1 shows the movements out of deciles for similar schools between the 2001-2002 and 2002-2003 academic years. Decile 10 contains the most highly ranked schools, decile 1, the lowest ranked. At the extremes there is nowhere to move but toward the middle and approximately half the schools do so. The movements are less lopsided for deciles closer to the middle but it remains true that the schools moving toward the middle consistently outnumber those moving away from the middle. Regression to the mean!

Table 1
Movements Out of Deciles for Similar Schools, 2001-2002 to 2002-2003

Decile	Percent Up	Percent Same	Percent Down
10	0.0	54.9	45.1
9	21.5	24.8	53.7
8	28.7	22.4	49.0
7	31.0	16.5	52.5
6	39.0	16.3	44.6
5	44.9	16.2	38.9
4	50.0	14.6	35.4
3	50.2	21.3	28.5
2	51.3	24.9	23.8
1	49.2	50.8	0.0

A school's average test score is a measure of the level of achievement, not improvement. While it is certainly a laudable goal to have every student achieve some level of proficiency, should some schools be rewarded and others penalized for factors beyond their control using a statistic that does not actually measure how much students learn during school year?

This argument suggests that an authentic measure of educational value added should look not at a school's average test score, but at changes in test scores over time. However, if statistical regression is not taken into account, changes in tests scores may be misinterpreted as changes in ability rather than fluctuations in scores about ability. These two issues—the desire to assess changes in abilities and the need to account for regression—suggest a different way of using test scores to evaluate schools: Kelley's equation.

Kelley's equation predicts the magnitude of the drop-off in above-average scores and the size of the improvement in below-average scores. Then we can see whether the observed regression is larger or smaller than that predicted by purely statistical arguments.

For the California 2001-2002 API scores, 4,749 schools scored above their estimated ability and 2,442 scored below, evidence that abilities did increase from the previous school year. Of course, as long as observed scores are an imperfect measure of ability, some schools will appear to be more successful and some less successful than they really are, which cautions against overreacting to a single year's scores. Evidence accumulated over several years is likely to be more reliable than evidence from a single year. The important point is that we should be looking at the right evidence—using Kelley's equation to take regression into account.

What Works and Doesn't Work

Test scores are used not only to evaluate schools, but also to compare different educational approaches, curricula, and teachers. Unfortunately, an insufficient appreciation of regression has caused a variety of errors in interpreting test scores.

Special Tutoring

Many schools use tests to identify students who need special tutoring, and then measure the success of the program by seeing whether these low-scoring students' do better after the tutoring sessions. Often overlooked is the fact that, because tests are an imperfect measure of ability, the students who are given special tutoring are generally not as weak as their initial low scores suggest. They can consequently be expected to do better on subsequent tests even if the tutor does nothing more than wave a hand over their heads.

We can use Kelley's equation to estimate the anticipated improvement in test scores due simply to statistical regression. Suppose that the overall average score on the test is 500 and we give tutoring to a group of students who averaged 300. If the test has a reliability of 90 percent, we estimate the average ability of the students in the low-scoring group to be 320. If the students receive special tutoring between the tests, their average score would have to exceed 320 on the second test to provide evidence that the tutoring made a difference.

This is not an argument for complacency. Those students with the lowest scores may not be as weak as they seem, but they still are weak students. They need help to increase their abilities. The regression argument is, in fact, an argument against complacency. An increase in test scores from 300 to 320 is no cause for celebration. It is a warning that whatever help these students received did no good at all.

SAT-Prep Classes

Many high school students take the SATs and are disappointed in their test scores. They enroll in expensive SAT-prep classes and, afterward, retake the SAT and get higher scores—apparent evidence that their money was well spent. What they haven't considered is that their initial low scores may have been below their ability—which is especially likely if they are disappointed in their scores. They can consequently expect, on average, to do better on the retake, whether or not they take an SAT-prep class.

On the other side of the SAT spectrum, a Harvard study of incoming freshmen found that students who had taken SAT preparation courses had lower SAT scores, on average, than did freshmen who had not taken such courses. Harvard's admissions director presented these results at a regional meeting of the College Board, suggesting that such courses are ineffective and "the coaching industry is playing on parental uncertainty." Why is this study unpersuasive?

It is surprising that Harvard's admission director would not be aware of likely differences between students who choose to take SAT-prep courses and those who do not. Those students who take SAT-prep courses are likely to feel they did not do as well as they had anticipated on their first attempt. Those who choose not to take an SAT-prep class are more likely to be satisfied with their scores, perhaps because they did better than they anticipated on their first attempt. If so, their ability is most likely lower than their SAT scores would indicate.

The Brightest Get Dimmer

When students with high test scores are retested, there is usually regression to the mean. Howard Wainer gave this example, which occurred when he was a statistician with the Educational Testing Service:

> My phone rang just before Thanksgiving. On the other end was Leona Thurstone; she is involved in program evaluation and planning for the Akebono School (a private school) in Honolulu. Ms. Thurstone explained that the school was being criticized by one of the trustees because the school's first graders who finish at or beyond the 90th percentile nationally in reading slip to the 70th percentile by 4th grade. This was viewed as a failure in their education. Ms. Thurstone asked if I knew of any longitudinal research in reading with this age group that might shed light on this problem and so aid them in solving it. I suggested that it might be informative to examine the heights of the tallest first graders when they reached fourth grade. She politely responded that I was not being helpful.

Wainer's helpful hint regarding heights was a reference to Galton's study of regression—a reference Ms. Thurstone did not understand because she was not aware of the regression problem.

Similarly, I once was on a committee interviewing candidates for a tenure-track academic position. One candidate, whose avowed specialty was educational testing, was asked how she would interpret data showing that a student who had scored one standard deviation above the mean on a test administered at the start of the school year scored 0.8 standard deviations above the mean on a similar test administered at the end of the year. Her answer was that the school had failed this student.

Kelley's equation comes to our aid again. Suppose that we identify a group of students who averaged 700 on a test where the overall average score is 500. To make this example especially provocative, suppose that the school tries to nourish these gifted students by putting them in special classes with an enriched curriculum and inspiring teachers. After the term ends, the students are retested and their average score has fallen from 700 to 690. Apparently, the special treatment not only did not help, it hurt.

Before we abandon this experiment, consider regression to the mean. If the test's reliability is 0.9, then, based on the first test, our estimate of their ability is 680. If the average score on the second test were 680, this would be nothing special. It would be just what we expected, whether or not the students are given a more challenging curriculum and stimulating teachers. If the average score falls from 700 to 690, this is not evidence that they were hurt by the special attention. A score of 690 is 10 points better than expected and indicates that their ability improved between the tests.

The Worst of Both Worlds

There can be mass confusion if we look at both the low-scorers and the high-scorers, and ignore regression. A study might evaluate a new teaching method by giving students a test (the pre-test) before using the new teaching method, and then giving a second test (the post-test) afterward. Suppose that the teaching method is, in fact, worthless and there is no change in the average student score. However, a careful examination of the data (as in Figure 2) reveals that the students with the highest scores on the pre-test did somewhat worse after the new teaching method while the lowest scoring students did somewhat better. The new teaching method seems to be good for weak students, but bad for good students.

Figure 2
A New Teaching Method is Good for Weak Students, But Bad for Strong Students

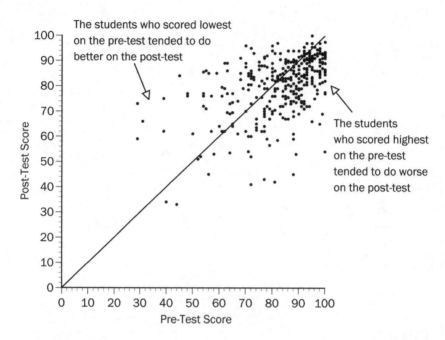

But what if (as in Figure 3) we happen to look at the students who got the highest scores on the post-test and find that they did somewhat worse on the pre-test, and that the lowest scoring students on the post-test did better on the pre-test? Now, the apparent conclusion is that the new teaching method works for good students, but is bad for weak students.

Either way, we would have been fooled by regression. There was no new teaching method behind the data in these two figures. Bonus points if you recognized that these graphs are just a recycling of the midterm and final examination scores in Chapter 4 (Figures 5 and 6). As with any two tests—midterm and final, pre-test and post-test, whichever and whatever—that are imperfectly correlated, the people with the highest and lowest scores on either test will tend to be closer to the mean on the other test. This statistical pattern tells us nothing about the effects of a new teaching method, or anything else that happened between the two tests.

Figure 3
A New Teaching Method is Good for Strong Students, But Bad for Weak Students

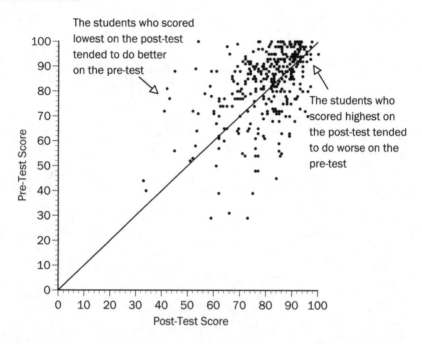

Another variation is to identify the students who improved the most between the pre-test and post-test (there will be some, as long as luck is involved) and then look for common traits (there will be some). Maybe it turns out that they mainly wore light-colored clothes; so you conclude that light-colored clothes works wonders. Or maybe most of them forgot to brush their teeth that morning; so you conclude that toothbrushing makes you dumber. Whatever theory you come up with—no matter how silly or plausible—the inevitable regression in the data is misleading evidence that supports your theory.

Strivers

I once attended a meeting where a black female executive said that she was not embarrassed that she had been hired and promoted over more qualified candidates, because it had been hard for her growing up black in Beverly Hills. An older white man whose parents had been Dust Bowl refugees from Oklahoma quietly replied, "If you think that was hard, try growing up in a tent in the cotton fields of Bakersfield."

Affirmative action policies have many goals, one of which is to level the playing field for those who have been disadvantaged in various ways. As far as socioeconomic disadvantages go, race is, at best, a noisy proxy. Every black person is not impoverished, and every white person is not wealthy.

In the 1990s the Educational Testing Service (ETS) commissioned a project that was called Strivers. The idea was that college admission officers could use a scientifically valid way of taking into account the socioeconomic background of applicants. The hope was that the program would increase the acceptance rates for African Americans and Hispanics in a way that was protected from legal challenges to race-based admissions policies.

The ETS study envisioned estimating the expected value of every student's SAT test score, based on fourteen socioeconomic factors, including parental education and income, the number of books in the home, the geographic location of the school (urban, suburban, or rural), and the quality of the high school the student attended. Students whose SAT scores were more than 200 points above their expected values were labeled "strivers" and, it was hoped, would be looked at favorably by college admission officers.

Anthony Carnevale, the project director, argued that, "A combined score of 1000 on the SATs is not always a 1000. When you look at a Striver who gets a 1000, you're looking at someone who really performs at a 1200. This is a way of measuring not just where students are, but how far they've come."

The project sounds sensible, but there were serious problems. One is that if college affirmative action programs were replaced by the Strivers program, the acceptance rates for African Americans and Hispanics would fall. Theodore M. Shaw, the president of the NAACP Legal Defense and Educational Fund noted that many middle-income and upper-income black students who were currently being admitted to selective colleges might not be admitted under the Strivers program, and they might be replaced by lower-income white students because, "there are more poor white students in this country than there are poor black students." Carnevale reportedly found that the Strivers model would reduce acceptance rates for African Americans and Hispanics unless race were explicitly included in the model—which would open the Strivers model to legal challenges.

A more subtle problem is that for those students who score above their expected value, their SAT score is more likely to be an overestimate of their ability than an underestimate. Let's apply Kelley's equation to Carnevale's example. At the time, SAT scores ranged from 400 to 1600, with an (approximate) mean of 1000, and a reliability of 0.9. If we knew nothing at all about a student, the estimated ability of a student who scores 1000 is 1000.

Now suppose that we know from the 14-factor Strivers model that the expected value of the student's score is 800. A score of 1000 marks this student as a Striver, and Carnevale argued that if a student who is expected to score 800 and scores 1000, "you're looking at someone who really performs at a 1200."

Wrong. Kelley's equation estimates this person's ability to be 980, which is *less* than 1000. A student who scores above his or her expected value probably has an ability *below* the score, not 200 points above it!

The prominent sociologist, Nathan Glazer, endorsed the Strivers project, arguing that, "It stands to reason that a student from a materially and educationally impoverished environment who does fairly well on the SAT and better than other students who come from a similar environment is probably stronger than the unadjusted score indicates." In fact, Kelley's equation implies the exact opposite. If an advantaged student and disadvantaged student both score 1000, the advantaged student probably has more ability, since Kelley's equation places the ability between the score and the expected value—down for a disadvantaged student, up for an advantaged student.

The underlying purpose of the SAT is to predict student success in college. Counting an SAT score of 1000 as 1200 does not do the student any favors if the student's ability is 980. Howard Wainer, who had been Principal Research Scientist at ETS for 21 years and is currently Distinguished Research Scientist at the National Board of Medical Examiners, wrote that,

This result should be distressing for those who argue that standard admission test scores are unfair to students coming from groups whose performance on such admission tests is considerably lower than average . . . and that they under-predict the subsequent performance of such students. Exactly the opposite of that is, in fact true, for there is ample evidence that students from groups who are admitted with lower than usual credentials on average, do worse than expected. The extent to which they perform more poorly is almost entirely predictable from Kelley's equation.

We don't know the specific reasons the ETS found compelling, but they pulled the plug on the Strivers project.

IV. GAMES OF CHANCE

8

Hopes and Excuses

EXTRASENSORY PERCEPTION (ESP) IS THE ABILITY TO RECEIVE INFORmation without using the five physical senses: sight, hearing, taste, smell, and touch. ESP includes both telepathy (reading another person's mind) and clairvoyance (identifying an unseen object). An example of telepathy is when one person (the sender) thinks of a number and another person tries to guess it. An example of clairvoyance is when a person tries to guess what is inside a gift package.

Every semester, I do a test of extrasensory perception (ESP) in my statistics classes. I flip a coin ten times and ask the students to attempt to read my mind and record their best guesses. To encourage serious concentration, I offer a one-pound box of chocolates from a local gourmet chocolate store as a prize. Figure 1 shows the results for one class of 29 students.

One student got eight correct; even more remarkably, another student got nine wrong. Believe it or don't, J. B. Rhine, the most famous ESP researcher in history, would interpret the student who got nine wrong as evidence of negative ESP (or "avoidance of target"). Some of his subjects got high scores and some got low scores. Rhine cited both as evidence of ESP. According to his creative interpretation, the high scorers had ESP and were trying to do well; the low scorers also had ESP but were deliberately giving the wrong answers in an effort to embarrass Rhine. If you want to believe something hard enough, you can find reasons to believe.

Figure 1
Students Guessing Coin Flips

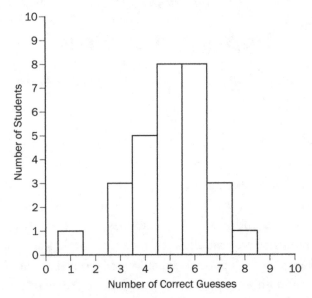

Rhine retested many of the high scorers, but their scores almost always declined. This happened so often that he called it the "decline effect." His explanation: "They were no doubt fatigued and bored." An alternative explanation is that the initial high scores were just lucky guesses and their subsequent scores regressed to the mean.

Suppose that there is no such thing as ESP, so that those who get unusually high or low scores are just lucky or unlucky. What is your best prediction of the number of correct guesses on a second test with ten new coin flips? The most likely outcome is five correct, though it is far from guaranteed. In fact, there is only about a 25 percent chance of exactly five correct and five incorrect. However, it is very likely (95 percent certain) that the person who got eight right will do worse the next time and even more likely (99 percent probability) that the person who got nine wrong will do better the next time. Regression to the mean teaches us that extraordinary performances (like eight right or nine wrong) are most likely going to be followed by less extraordinary performances. We shouldn't be surprised and we need not invent excuses like the decline effect.

This would be true even if ESP were real (as long as no one gets

100 percent right every time). Someone who has modest ESP with a 60 percent chance of giving a correct answer would average 60 percent correct in the long run. Even if this person were lucky enough to get eight out of 10 right, we can expect that she will average 60 percent correct going forward, which is again regression to the mean.

If my students can't read my mind, then my ESP test is a pure game of chance. They might as well flip their own coins and see how often their flips match mine. At the other extreme are games that are pure skill in that the outcome is the same every time. In Tic Tac Toe, two players who understand the game will tie every game with no variation whatsoever in the outcome. In a foot race between a fast grownup and a slow child, the grownup will win every time. There is no luck, because there is no uncertainty about the outcome. There is no regression, because the outcome is the same every time.

In between games that are all-chance and those that are all-skill, are games that are a mixture of chance and skill; for example, poker. Some poker players are better than others and will win more often in the long run; however, the outcome of any single hand is uncertain, in part because of the proverbial luck of the draw.

Suppose that when two people ("Ace" and "King") play heads-up poker, just one person against the other, Ace can expect to win, on average, $2 per hand in the long run. In the first several hands, Ace may win an average of $5 per hand. We can anticipate Ace regressing to the mean with his future wins averaging $2 per hand. Similarly, if Ace has been losing $4 per hand, we can expect him to regress to the mean by winning an average of $2 per hand going forward.

Some players don't think this way. Some think that winning makes winning more likely. Other believe that winning makes losing more likely. Both are wrong.

Hot and Cold Streaks

In his final newspaper column, Melvin Durslag, a member of the National Sportscasters and Sportswriters Hall of Fame, reminisced about some of the advice he had received in his 51 years as a sports columnist, including this suggestion from a famous gambler: "Nick the

Greek tipped his secret. He trained himself so that he could stand at the table eight hours at a time without going to the washroom. It was Nick's theory that one in action shouldn't lose the continuity of the dice."

In pure games of chance involving coins, dice, and roulette wheels, each outcome is independent of other outcomes, past, present, or future. In any fair game, a player will win some and lose some (or, it often seems, win some and lose many). The wins are at times scattered and, at other times, bunched together. Some gamblers mistakenly attach a great deal of significance to these coincidental clusters. They apparently believe that luck is an infectious disease that a player catches and then takes awhile to get over. For example, Clement Mc-Quaid, "author, vintner, home gardener, and keen student of gambling games, most of which he has played profitably," offers this advice:

> There is only one way to show a profit. Bet light on your losses and heavy on your wins. Many good gamblers follow a specific procedure: a. Bet minimums when you're losing. . .b. Bet heavy when you're winning. . . c. Quit on a losing streak, not a winning streak. While the law of mathematic probability averages out, it doesn't operate on a set pattern. Wins and losses go in streaks more often than they alternate. If you've had a good winning streak and a loss follows it, bet minimums long enough to see whether or not another winning streak is coming up. If it isn't, quit while you're still ahead.

Aha! Good idea. You will indeed show a profit if you win your large bets and lose your small ones, but how do you know in advance whether you are going to win or lose your next bet? Suppose you are playing a dice game and have won three times in a row. You know that you have been winning and you are excited, but dice have no memories or emotions. Games were invented by people. Dice don't know the difference between a winning number and a losing number. Dice do not know what happened on the last roll and do not care what happens on the next roll. The outcomes are independent in that the probabilities are constant, roll after roll.

Yet, gamblers, hoping to find a way to beat the odds, cling to fanciful illusions of hot and could streaks. One recommended this strategy for winning at craps at Las Vegas:

> [F]ind a hot table. Never remain at a cold one. Always make it a policy to keep looking—move around! A tip-off might be the yelling crowd where a hot roll may be taking place. Another indication is a lot of money spread every which way around the table by numerous players. . . . [I]t is far more lucrative to tag along on the tail end of a hot roll than to go in fresh on a cold one. And no one moving from table to table will actually catch a hot roll from the beginning. If 65 percent of a streak is caught, it's enough!

People who win a dice game several times in a row want to believe that they are on a hot streak and will keep winning. If the hot streak continues, true believers are even more convinced that they are riding a hot streak. If it doesn't continue, they invent fanciful excuses so that they can cling to their nonsensical theory.

It's not just games of chance played with coins, dice, and roulette wheels. In the second game of the National Basketball Association (NBA) finals between long-time rivals Boston Celtics and Los Angeles Lakers, Celtic guard Ray Allen made seven three-point shots in a row. One teammate said it was "Incredible." Another said it was "Unbelievable." One sportswriter wrote that, "Allen got hot." Another wrote that Allen had "slipped into that shooting zone only visited by real-life superstars and movie characters."

Ray Allen isn't unique. Many basketball players have made (or missed) several shots in a row. Many football quarterbacks have thrown several consecutive complete (or incomplete) passes. Many baseball players have made several hits (or outs) in a row. Fans and players alike see these streaks and conclude that the player has gotten hot or cold. Purvis Short, who averaged 17 points a game over his 12-year National Basketball Association career and once scored 59 points in a single game, expressed the common perception: "You're in a world all your own. It's hard to describe. But the basket seems to be so wide.

No matter what you do, you know the ball is going to go in."

We see a pattern and hatch a theory to explain the pattern. If a basketball player makes several shots in a row, it must be because he is hot, with an increased probability of making shots. If a player misses several shots in a row, it must be because he is cold, with a reduced probability of making shots. What most people—fans and players—do not appreciate is that even if each shot, pass, or swing is independent of previous shots, passes, and swings, coincidental streaks can happen by chance alone.

If a fair coin is flipped ten times, there is a 47 percent chance of a streak or at least four heads in a row or four tails in a row. If the coin is flipped 20 times, there is a 77 percent chance of such a streak. Of course, these are coin flips, not real basketball shots. But that is exactly the point! Hot and cold streaks often appear, simply by coincidence, in completely random coin flips. This does not prove that athletic hot and cold streaks are just coincidences. However, it does caution that a streak of several successes in a row does not ensure continued success, nor does a streak of several failures guarantee continued failure. Hot and cold streaks may be nothing more than luck.

Ray Allen took more than 7,000 three-point shots in his NBA career and made 40 percent of them. Imagine a coin with a 40 percent chance of heads being flipped 7,000 times. It is almost certain that there will be a streak of seven heads in a row at some point in those 7,000 flips. Ray Allen's seven-shot streak may be no more meaningful than seven heads in a row in the midst of 7,000 coin flips.

The Gambler's Fallacy

Some believe in the opposite—that a streak of good luck makes bad luck more likely. For example, one gambler wrote that,

> Regardless of mathematics and the theory of probability, if I'm in a craps game and the shooter makes ten consecutive [wins], I'm going to bet against him on his eleventh throw. And if he wins, I'll bet against him again on the twelfth. Maybe it's true that the mathematical odds on

every throw of the dice remain the same, regardless of what's gone before—but how often do you see a crap-shooter make 13 or 14 straight [wins]?

Thirteen consecutive wins is rare, but it is a very different question to ask the probability of 13 straight wins, given that you have 12 wins already. The outcome of a fair game of chance does not depend on what happened in the past. Yes, good luck doesn't last forever. That's why it is called *luck*. You will have bad luck eventually, but good luck does not make bad luck more or less likely.

The belief that every bit of good fortune makes bad fortune more likely, and that every spell of bad luck makes good luck more likely is called the gambler's fallacy, or the fallacious law of averages. In fair games, the chances of good luck and bad luck don't change simply because there hasn't been any recently.

Here is a good example of someone addicted to the gambler's fallacy:

This man has kept records of play on Blackjack, craps, and Roulette, and the house percentage on all three games works out inexorably, within a fraction of a percentage point—but there are times shown by his records when there are lengthy streaks of steady house wins along with other streaks of house losses.

His records show that such streaks exist. But no matter how long he studies his figures, he hasn't been able to arrive at any pattern in them. He can see that the streaks are there, but he hasn't anything even resembling an explanation of why they occur when they do or why they last for a certain period.

"Nonetheless," he insists, "if I'm scoring red and black in a Roulette game and red has come up 500 times in 900 spins of the wheel, I'm going to put my money on black the next 100 spins. What's more, I'll bet you that I come out ahead of the game."

The only good thing I can say about this hopeless optimism is that the more time he spends studying the numbers, the less time he spends betting on them.

In honest games, betting strategies based on the law of averages do not work. In fact, the few gambling systems that do work are based on the opposite principle—that physical defects in the apparatus cause some numbers to come up more often than others. In the late 1800s, an English engineer, William Jaggers, hired six assistants to spend a month recording the winning numbers on Monte Carlo roulette wheels. Instead of betting against these numbers, counting on things to average out, he bet on these numbers, counting on wheel imperfections to continue causing these numbers to win more often than others. He won nearly $125,000—more than $6 million in today's dollars—until the casino caught on and began switching wheels nightly.

Many people who buy lottery tickets believe in the gambler's fallacy. A study of the numbers chosen for lotto games found that ticket buyers tend to avoid numbers that have won recently, evidently because they believe that the more often something has happened in the past, the less likely it is to happen in the future.

Many sport fans are addicted to the gambler's fallacy. During an NFC championship game between Washington and San Francisco, it appeared that the game might be decided by a field goal. One of the CBS commentators, Jack Buck, pointed out ominously that Mark Moseley had already missed four field goals. Hank Stram, the other CBS commentator, responded, "The percentages are in his favor." I'd say that the percentages are that he is going to lose his job if he doesn't start making field goals.

I once watched a kicker miss three field goals and an extra point in an early-season college football game. The television commentator said that the coach should be happy about the misses as he looked forward to some tough games in coming weeks. The commentator said that every kicker is going to miss some over the course of the season and it is good to get these misses "out of the way" early in the year. I'd say the coach should start thinking about a new kicker.

I once heard a Little League father tell his son to lend his baseball bat to a poor hitter "to use up the bat's outs." A Major League Baseball manager once put in a pinch hitter for Hall of Fame pitcher

Ted Lyons because he had made four hits in a row and the manager reasoned that pitchers never get five hits in a row. Well, they certainly won't if they never have an opportunity to bat after four hits in a row.

At the midpoint of the Cape Cod League baseball season, Chatham was in first place with a record of eighteen wins, ten losses, and one tie. (Ties are possible because games are sometimes stopped due to darkness or fog.) The Brewster coach, whose team had a record of 14 wins and 14 losses, said that his team was in a better position that Chatham: "If you're winning right now, you should be worried. Every team goes through slumps and streaks. It's good that we're getting [our slump] out of the way right now." Maybe that was a lame attempt at a pep talk, but it makes no sense. Losing baseball games does not automatically make winning more likely. If anything, the team probably isn't very good. The team with a better record is probably a better team and is certainly in a better position to win the championship, which Chatham did.

In 1989, before pitching a baseball game for the Los Angeles Dodgers, who had not scored a run in 24 consecutive innings, Orel Hershiser said, "I prefer pitching on the day after we've been shut out because I figure we're going to score. I guess the odds are really in my favor now." Maybe he was joking. Hershiser had a terrific earned run average of 2.31 that year, but only went 15-15. He was winless in one nine-game streak where the Dodgers scored a total of nine runs.

Unlike coins and roulette wheels, humans have memories and care about wins and losses. Still, the probability of making a field goal doesn't go up simply because a kicker has missed four in a row. The chances of a base hit in baseball do not increase just because a player has not had one lately.

One of the most successful stock market investors of all time is Warren Buffett, who has made about 25 percent a year over some fifty years, while the overall stock market has averaged closer to 10 percent a year. A 2013 *Bloomberg Businessweek* article noted that Warren Buffett's Berkshire Hathaway had underperformed the S&P500 for four months in a row, June, July, August, and September, 2013, and concluded that "Berkshire stock is due for a comeback vs. the S&P." This, too, is the fallacious law of averages. Doing worse than the market averages does not increase the chances that one will do better than the market. If anything, it suggests that Buffett may have lost his Midas touch.

Would you be comforted by a doctor who said, "Five out of ten people who have this disease die of it. It's lucky you came to me; my last five patients all died." Or how about this one: "My next marriage is sure to work; the odds against five divorces are astronomical."

When we are having a run of bad luck, we hope that we are due for a change in fortune. It *is* unlikely that our bad luck will continue forever, but every failure does not make success more likely, or vice versa. We usually have to change our behavior to change our fortune. A doctor whose patients are dying needs to rethink his practice. A person who has been divorced four times needs to rethink her behavior and choice of husbands.

Which is it, hot streak or the law of averages? Jimmy the Greek, a celebrity sportscaster and Las Vegas bookmaker, said that, "The smart professional gambler, when heads comes up four times in a row, will bet that it comes up again. . . . The amateur bettor will figure that heads can't come up again, that tails is 'due'." The truth is that both are wrong. Success will regress to the mean in that future successes will be consistent with the probability of success, but that probability does not change—up or down—because of past successes or failures.

Don't Look Back

I have some friends (the "Nesterovs") who had a 12-foot by 16-foot problem area in their backyard. It was too small for a patio, too big for a path, and didn't get enough sun to grow anything they were interested in planting. While visiting a home-supply store, they noticed some 16-by-24-inch cinder blocks with small diamond-shaped holes that could be used for growing ground cover. And, best of all, they were cheap!

So, the Nesterovs laid down 72 cinder blocks and put ground cover in the holes and tried to remember to water. Unfortunately, they didn't remember often enough and the ground cover died. Now they had a 12-by-16-foot grid of cinder blocks with diamond-shaped holes—what they laughingly called "ankle breakers."

Figure 2
Ankle Breakers

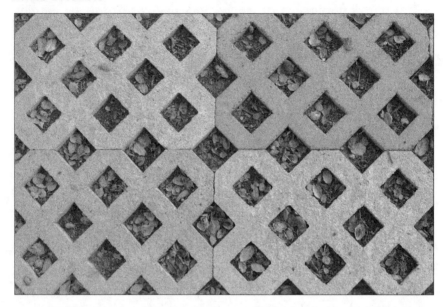

Clearly, these cinder blocks were a bad idea. But the Nesterovs had bought them and what else were they going to do with 72 ankle breakers? They figured that they would have to use them for at least 10 years to "get their money's worth," and that is what they did.

These ankle breakers are a sunk cost because once the money is spent, it cannot be unspent. Whether it was spent wisely or foolishly is unimportant. The relevant question is not whether they should have bought cinder blocks (they shouldn't have), but whether they should replace them with something more attractive and less dangerous. The relevant cost is not how much they paid for their 72 ankle breakers, but how much they would have to pay to put in something better.

It's human nature to feel that we must cover up our mistakes. You buy a colossal ice cream sundae for a special price but, halfway through, you're feeling sick. Do you finish the sundae because you paid for it? You have season tickets to college football games in the Midwest. Come November, the team sucks and the weather is worse. Do you go to the game because you paid for the tickets? You buy a stock right before some unexpected bad news about the company causes the price to fall. Do you sell for a loss to get the tax benefit or do you hold onto the stock because selling it for a loss is an admission that you made a mistake buying the stock in the first place?

There is nothing to be gained and much to be lost by moping about things you can't change. Things that can't be changed are called sunk costs. The Nesterov's ankle breakers are a sunk cost. The ice cream sundae you bought but would get sick finishing is a sunk cost. So are tickets to a bad football game. So is your unfortunate stock purchase.

In games of chance, losses are losses and cannot be undone, yet many people behave differently after losses. Kahneman and Tversky observed that bets on long shots at horse races increase towards the end of the day, because people want a cheap way to win back what they lost earlier in the day. They concluded that a "person who has not made peace with his losses is likely to accept gambles that would be unacceptable to him otherwise."

Two students and I looked at this question by analyzing hundreds of thousands of Texas Hold 'Em poker hands at an online poker site with blinds (initial bets) of $25 to $50, which are considered high-stakes tables and attract experienced poker players. We considered a hand where a player won or lost $1,000 to be a significant win or loss. After a big win or loss, we monitored the player's behavior during the next 12 hands—two cycles around a six-player table. We followed two cycles because experienced players often make no voluntary bets, and 12 hands are still reasonably close to the time of the big win or loss. We restricted our analysis to individuals who played at least 50 hands in 12-hand windows following their big wins and at least 50 hands in 12-hand windows following their big losses.

Our final data set included 346 players who met the various criteria. The median number of hands played was 1,738, with half of the players playing between 717 and 4,549 hands. Half of the players won or lost more than $200,000, ten percent won or lost more than $1 million.

At the beginning of each hand, the player sitting directly to the left of the dealer puts a small blind of $25 into the pot, and the next player puts in a big blind of $50. Each player is then dealt two "hole cards" that only they see. The players who have not already put money in the pot decide whether to play or fold. To play, the players must either "call" the big blind, $50, or raise the bet above $50, forcing the other players to match the highest bet on the table. The bets go clockwise around the table until the highest bet is called by all players who wish to remain in the hand, or all but one person folds.

If more than one player is still in, three community cards ("the flop") are dealt, which are visible to everyone and can be used by each player to build the best possible hand. Another round of betting occurs, starting with the person to the left of the dealer. After this round of betting, a fourth community card ("the turn") is dealt, and there is another round of betting. Finally, the fifth community card ("the river") is dealt, and there is a final round of betting. The player with the best five-card hand, between their two cards and the five community cards, wins the pot.

We looked at games played at either six-player tables or two-player ("heads up") tables. There are sometimes empty seats at a six-player table, and poker strategy depends on the number of players at a table; for example, the chances that a pair of eights in the hole will yield the best hand declines as the number of players increases. We consequently grouped the data for six-player tables according to the number of players at the table. We did not combine the data for heads-up tables with the data for six-player tables with two players because the people who choose to play heads-up poker may have different styles than players who choose a six-player table but occasionally have four empty seats.

The generally accepted measure of *looseness* is the percentage of hands in which a player voluntarily puts money in the pot. This can include a call or a raise, but does not include blind bets since these are involuntary. After a hand is dealt, everyone other than the player who put in the big blind must either bet or fold before they see the three-card flop. Thus, looseness measures how often a person puts money into the pot in order to see the flop cards.

Tight players fold when their two hole cards are not strong; loose players stay in, hoping that a lucky flop will strengthen their hand. At six-person tables, people are typically considered to be very tight players if their looseness is below 20 percent and to be extremely loose players if their looseness is above 50 percent. For our data set, the average looseness values range from 51 percent at heads-up tables to 26 percent at full six-player tables.

In theory, experienced poker players have a style that they feel works for them based on hundreds of thousands of hands they have played over years or decades. Once they have settled on their optimal strategy, they should stick to it, no matter what the outcome of the last few hands. If they suffer a big loss, they should recognize that it was bad luck and stick to their strategy, believing that their ability will win back the money they lost eventually. Their success will regress to the mean in that the big loss was an aberration and future hands will generate profits consistent with their ability.

We found that, in practice, players typically changed their style of play after winning or losing a big pot—most notably, playing less cautiously after a big loss, evidently hoping for lucky cards that will erase their loss quickly.

Table 1 shows that it is consistently the case that more players are looser after a big loss than after a big win; for example, with six players at the table, 135 players were looser after a big loss than after a big win, while the reverse was true for only 68 players. To test the robustness of this conclusion, we also looked at $250 and $500 thresholds for a big win or loss and found that in every case, most players play looser after a large loss. We also found that larger losses are more memorable in that the fraction of the players who played looser increased as the size of the loss increased.

Table 1 Player Looseness

Players at Table	Number of Players	Average Looseness	Players Looser After	
			Big Win	Big Loss
heads-up	228	51	74	154
2	40	46	17	23
3	33	35	11	22
4	75	29	21	54
5	150	26	53	97
6	203	26	68	135

Was this change in strategy profitable? If experienced players are using profitable strategies to begin with, changing strategies is a mistake. That's exactly what we found. Those players who played looser after a big loss, did worse than they normally did.

Despite having played many, many poker hands, these experienced players did not recognize that their performance would regress to the mean after a big loss. Instead, they felt compelled to play more speculatively, hoping to cover their losses quickly.

This behavior may be applicable to other decisions. Ken Warren, a well-respected poker writer, once said about Texas Hold 'Em: "More money is lost by players who know what the right thing to do is, but don't do it, than for any other reason. Having a strategy, a game plan and the discipline to stick to it are, along with a sufficient bankroll, the four most important things that a player needs to be a winner." David Nelson, the Senior Vice President of Legg Mason Funds, wrote about Warren's observation: "You could say the same thing about investing. Game plan, strategy, discipline and obviously, bankroll."

If investors are like poker players, their behavior might well be affected by large losses; for example, making otherwise imprudent long-shot investments with the hope of offsetting prior losses cheaply. Various studies have found that:

1. Treasury bond traders are much more likely to take large risks in the afternoon if they have morning losses.
2. Floor traders on the Chicago Mercantile Exchange increase their risk exposure after losses.
3. Professional stock day traders who lose money in the morning trade more aggressively in the afternoon.
4. Mutual funds and portfolio managers who are not performing up to their target levels take greater risks to try to hit their targets.
5. Several catastrophic wagers placed by "rogue" traders were desperate attempts to cover earlier losses.

A 2009 article in the *Wall Street Journal* reported that many investors were reacting to their stock market losses by making increasingly risky investments:

The financial equivalent of a "Hail Mary pass"—the desperate attempt, far from the goal line and late in a losing game, to fling the football as hard and as high as you can, hoping it will somehow come down for a score and wipe out your deficit.

For poker players and investors who have sound strategies, regression to the mean counsels that patience is better than a Hail Mary.

V. SPORTS

9

Champions Choke

WHEN MALAYSIAN BADMINTON PLAYER MOHD HAFIZ HASHIM won the 2003 All England Badminton Championship, arguably the world's most prestigious badminton tournament, the Malaysian Prime Minister, Mahathir Mohamad, said, "Very good and congratulations, but now I would like to request everybody not to spoil him. . . . I hope the states will not start giving acres of land and money in the millions, because they all seem not to be able to play badminton after that."

Sure enough, Hashim never won All England again. Was he spoiled by the attention and largesse that accompanied his stunning victory? It's possible, but a more plausible explanation is that, in addition to their ability, tournament champions usually have luck on their side. If ability were all that mattered, the best player would win every point of every game. Luck does matter and top-flight badminton games often could go either way. Hashim won his semifinal match 15 to 12 and 15 to 12 and won the finals match 17 to 14 and 15 to 10.

Hashim never was as successful after the 2003 All England, but he was never as successful before the tournament either. The highest world ranking Hashim ever achieved was sixth. In the 2003 All England, he defeated several strong players, including the number one ranked player in the world in the championship match. Maybe the correct interpretation of Hashim's 2003 victory was that he was a very good player who, for one magical week, was very, very lucky. He regressed to the mean before and afterward.

Athletic contests may not seem to be games of chance, but they are in a sense if the outcome isn't known in advance. Even something as simple a 100-meter race between the fastest runners in the world has an element of chance if we are not certain who will win. Perhaps one runner gets a late start or slips slightly. Whatever the reason, as long as we don't know for sure who will win, as long as the outcome cannot be predicted with 100 percent accuracy, we can say that there is uncertainty; i.e., chance.

It is very hard to acknowledge the role of luck in athletic successes, even when it is right in front of our eyes. My college alumni magazine had an article about Murray Thompson, a remarkable graduate who, after working 39 years in the aerospace industry, decided to take up a sport called reined cow horse (or "working cow horse"), in which the horse and rider control a cow the way real vaqueros separate a cow from its herd or return it to the herd. At age 62, Thompson won a world championship. The article quoted a friend and fellow rider: "Murray's focus was phenomenal, and he peaked at exactly the right time." This friend didn't think it was luck; Thompson chose to focus and peak at the right time.

Yet, the very next paragraph quoted Thompson:

It's basically a miracle if you win. In the most recent competition in Reno, a pro rider on one of the best horses I saw was having a spectacular run, and at the last second, the cow they were working just ran over them. Two years of training and in the last few seconds it all falls apart. It was beyond the control of both horse and rider; they just got a bad cow, as they call it. There is so much that can go wrong, and to win it, all has to go right.

Precisely. Thompson was trying to be modest, but he got it exactly right. The importance of luck is right there in front of our eyes, but we seldom see it.

If somebody wins a championship, we think it is because they trained hard, focused, and peaked at the right time. We don't want to admit that part of the reason our heroes win is that they were lucky.

It's not just badminton players and cowboys. Olympic gold medalists often do not perform as well after the Olympics. Figuring there must be a reason, commentators conclude that the gold medalists consciously raised their ability in order to win the gold medal and then relaxed and let their ability fall afterward. Maybe. But perhaps the reason they won gold medals is that their luck peaked at the right time, and luck is something they have no control over.

It is not just individual athletes. It is teams, too. The year after the Seattle Seahawks won the 2014 Super Bowl, a *New York Times* article was titled, "Seahawks Battle Jinx of the Super Bowl Winner." The author noted that the last team to win the Super Bowl two years in a row was New England in 2004 and 2005. A player explained why he thinks it is so hard to repeat as champion: "When you go through the parade and get the ceremony and get the ring, it makes you forget how hard it was to get to that point, to reach the mountaintop." The Seattle coach agreed that, "It's been difficult for people, and the history of it shows you that it's hard to come back and get yourself back into this kind of position again." Jimmy Johnson, one of only six coaches to win back-to-back Super Bowls, said that the reason most teams don't repeat is that, "More times than not, complacency is what hurts a football team."

We want to believe that success is earned—that the team that wins the Super Bowl really is the best team and that, if it fails to repeat, it must be because it didn't work as hard as it did the year before. It was spoiled by success and, as a consequence, is no longer the best team.

Even though we are loathe to admit it, there is luck involved. When Seattle won the Super Bowl in 2014, its regular season record was 13-3, the same as the Denver Broncos. There were three teams that were 12-4, and four that were 11-5. With luck, any one of these nine teams could have been champion.

Seattle was lucky to even be in the Super Bowl. In the NFC Championship game, which would decide which team would play in the Super Bowl, Seattle squeaked by, 23 to 17, over the San Francisco 49ers, who turned the ball over three times in the fourth quarter (a fumble and two interceptions). If the Seahawks and 49ers had played each other ten times, Seattle would not have won every game. Seattle was lucky to win the game they did play.

As it turned out, Seattle made it back to the Super Bowl in 2015, giving it a chance for back-to-back championships, but it took a dramatic come-from-behind victory over the Green Bay Packers to get there. In a wild game filled with turnovers and dropped catches, not to mention a fake field goal, onside kick, and two-point conversion, Seattle was behind by twelve points with less than four minutes to go in the game, then rallied to tie the game and win it in overtime.

The Super Bowl itself was a classic that could have gone either way. One of Seattle's wide receivers, who had started the season working in a shoe store and had never caught a pass in the NFL, was Seattle's leading receiver, with four catches for 109 yards. It was that kind of game.

Late in the game, the New England Patriots, led by Tom Brady, scored what looked to be the game-winning touchdown. Then, with 48 seconds left in the game and 38 yards to go for a touchdown, the Seattle quarterback, Russell Wilson, threw a long pass to a well-covered receiver. The defender batted the ball away but as the Seattle player fell to the ground, the ball miraculously came down on his legs and bounced towards his arms, where he juggled the ball and then hung on for a catch, giving Seattle the ball first-and-ten on the five-yard line.

What if the receiver had not landed on his back? What if the batted ball had not landed on the fallen receiver's legs? What if the ball had not bounced off the receiver's legs toward his arms? What if his arms had not been in a position to catch the ball? What if the defender had landed in a position to knock the ball away again? So many what-ifs and could-have-beens! This play was miraculous because there was so much luck involved.

On the first down after the wonder catch, Seattle handed the ball to Marshawn Lynch ("Beast Mode") who had been giving the Patriots fits—averaging more than four yards a carry on 24 carries so far. Lynch gained four yards, taking the ball to the half-yard line. Surely, he would pound the ball into the end zone for the winning touchdown. Inexplicably, Seattle decided to pass the ball. The pass was intercepted and the Patriots were the winners.

This wasn't the only could-have-been moment in the game. If these teams had played ten times, who knows which team would have won more games. What we do know is that neither team would have won all ten games. This is why it is useful to acknowledge this uncertainty by labeling it luck. Luck doesn't mean that somebody flips a coin to determine the winner. Luck is just a convenient way of reminding us that football games—like most sporting events, like much of life—is uncertain.

In Major League Baseball (MLB) and the National Basketball Association (NBA), the championship is decided by a seven-game series. They don't do that in the NFL because football is an extremely violent game and it takes several days for the players to recover enough to play again. A professional football player once said that he knew it was time to retire when, after a Sunday game, he hadn't healed enough to play the following Sunday. Teams that are asked to play on Monday night (to generate television revenue) generally suffer the following Sunday because the aches and pains linger. If the NFL Championship were decided by a seven-game series, it might take seven weeks to determine a winner. So, they play one game (the Super Bowl) and pretend that one game is enough to determine the better team.

Even if one team has an 80 percent chance of beating the other team, there is still a 20 percent chance the weaker team will win. No matter which team wins, there is luck in that the other team could have won.

The National Football League (NFL) prides itself on "parity." They have a revenue-sharing system and other rules that make the teams more evenly matched than in baseball or basketball. There is even a "circle of parity" created by fans each year showing how each of the 32 teams beat another team, which beat another team, and so on, back to wherever you start in the circle. In 2014, one circle of parity showed that the Seahawks beat the Broncos, who beat the Cardinals, who beat the Raiders, who beat the Chiefs, who beat the Patriots, . . . who beat the Chargers, who beat the Seahawks.

There are several good teams in the NFL that, with a little luck, could win the Super Bowl. Each year, one of these good teams is lucky enough to win. The next year, another team may well be luckier. It is not completely random. In the next eleven Super Bowls after the Patriots repeated in 2005, there were several teams that played in the Super Bowl multiple times: New England three more times, Seattle three times, and Pittsburgh three times, Indianapolis twice, the New York Giants twice, and Denver twice. No team won back-to-back Super Bowls. (If we include 2002 and 2003, New England played in six of the fifteen Super Bowls between 2002 and 2016) The many repeat appearances demonstrate that ability does matter, but the dearth of back-to-back champions demonstrates that luck matters, too.

Here is a simple example. On every play, the offense calls a play based in part on what it expects the defense to do, while the defense calls a play based in part on what it expects the offense to do. Sometimes, the guesses are correct, sometime they are wrong—just like a student guessing the answer to a test question—and we can call the guesses that turn out to be correct good luck and the guesses that turn out to be wrong bad luck.

Fumble recoveries are another example of luck, in that no one can predict in advance which team will recover a fumbled football. A team can win a game because of a fumble recovery, but a fumble recovery this week has no effect on the team's chances of winning next week.

The Luck of the Packers

In case you aren't convinced yet, let me give another, more complicated example involving the January 11, 2015 playoff game between the Green Bay Packers and the Dallas Cowboys. The winner would go on to the NFC championship game, with a chance to play in the Super Bowl. The loser would go home to think about would-have-beens and could-have-beens. And there were a lot of would-have-beens and could-have-beens in this game.

One occurred with 4 minutes and 44 seconds to go in the game and the Packers leading 26 to 21. The Cowboys had the ball on the Packers 32-yard line, fourth down with two yards to go for a first down. Instead of going for a short-yardage play that would give the Cowboys a first down. The Dallas quarterback, Tony Romo, threw a long pass to the team's star receiver Dez Bryant. Bryant leaped above the Packer defender, grabbed the ball, and, as he came down, stretched towards the goal line, hoping to get a touchdown that would give the Cowboys the lead. He landed just short of the goal line, but it was apparently a first down for the Cowboys with about six inches to go for a touchdown. The nearest referee said it was a catch and the Cowboys had the ball, first and goal on the six-inch line. The Cowboys came up to the line of scrimmage, ready to score the go-ahead touchdown.

Then the Packers' coach threw his challenge flag, asking the officials to review the play from all camera angles. It seemed to most everybody that the Packers coach had inexplicably wasted a challenge (each team only gets two challenges in a game, plus a third challenge if the first two challenges are successful.) The official on the field had ruled it a catch and a ruling on the field is only overturned if the slow-motion replays prove conclusively that the ruling was incorrect. Here, everyone had seen Bryant's phenomenal catch clearly. What was the Packers coach thinking?

After reviewing the videos, the officials announced that the call was overturned. It was an incomplete pass and, since Dallas had failed to get a first down, Green Bay now had the ball, first down on its 32-yard line. The rule is that a pass receiver "must maintain control of the ball throughout the process of contacting the ground," and Bryant had apparently bobbled the ball briefly after hitting the ground. This is one of the most confusing and controversial rules in football, and experienced commentators often disagree about whether the officials made the right call. There was ability in Bryant's acrobatic catch, but there was also luck in the brief bobble, the Packers coach's decision to throw the flag, and the decision by the officials to call it an incomplete pass.

There was another memorable ability/luck tussle a few minutes later. Green Bay got a first down and was trying for another first down that would guarantee victory. There were now less than two minutes left in the game and Green Bay had the ball on third down with eleven yards to go for a first down. If Green Bay got a first down, they would be able to run out the clock. If they didn't, they would have to punt and Dallas would have another chance to win the game.

The Green Bay quarterback, Aaron Rodgers, tried to throw a pass to Randall Cobb, who was tightly covered by a Dallas defender. One of Dallas's defensive linemen tipped the pass after it left Rodgers' hand and the ball fluttered towards Cobb. Cobb happened to see the deflection, while the Dallas defender did not. Cobb dove for the ball and made a tremendous catch just before the ball hit the ground. It was an eleven-yard pass completion, and Green Bay had its first down and was able to run out the clock. There was ability in the great catch made by Cobb, but there was also a lot of luck in that the ball was tipped, and it happened to be tipped to a place where it could be caught, and that place happened to be just far enough to give Green Bay its crucial first down.

A championship team needs ability and luck. Ability will get a team into the playoffs and give it a shot at the title, but luck is needed to get to the Super Bowl and win the championship playing against other teams with lots of ability.

The *Sports Illustrated* Jinx

Many sports fans are convinced that champions choke—that athletes who achieve something exceptional usually have disappointing letdowns afterward. Evidently, people work extraordinarily hard to achieve extraordinary things, but once they are on top, their fear of failing causes the failure they fear.

The most famous example is the *Sports Illustrated* cover jinx. After Oklahoma won 47 straight college football games, *Sports Illustrated*'s cover story was, "Why Oklahoma is Unbeatable." Oklahoma lost its next game, 7 to 0, to Notre Dame. After this debacle, people started noticing that athletes who appear on the cover of *Sports Illustrated* are evidently jinxed in that they do not perform as well afterward. In 2002, *Sports Illustrated* ran a cover story on the jinx with a picture of a black cat and the wonderful caption "The Cover No One Would Pose For." More recently, we have the Madden Curse, which says that the football player whose picture appears on the cover of Madden NFL, a football video game, will not perform as well afterward.

The *Sports Illustrated* Jinx and the John Madden Curse are extreme examples of regression to the mean. When a player or team does something exceptional enough to earn a place on the cover of *Sports Illustrated* or *Madden NFL*, there is essentially nowhere to go but down. To the extent luck plays a role in athletic success, and it surely does, the player or team that stands above all the rest almost certainly benefitted from good luck—good health, fortunate bounces, and questionable officiating. Good luck cannot be counted on to continue indefinitely, and neither can exceptional success.

Football Regression

Chapter 1 discussed how Peyton Manning, who had the highest quarterback rating in 2013, regressed to the mean in 2014. It wasn't just Peyton. Regression is a general tendency, not a unique occurrence for one individual.

Figure 1 shows the quarterback ratings (QBRs) for all NFL quarterbacks who had at least 100 pass attempts in 2013 and 2014. Those quarterbacks above the 45-degree line did better in 2014 than in 2013; those below the line did worse in 2014 than in 2013. Most of the top-performing quarterbacks in 2013 did not do as well in 2014, while most of the low-performing quarterbacks in 2013 did somewhat better in 2014. There was a regression toward the mean in that those quarterbacks who were farthest from the mean in 2013 were closer to the mean in 2014.

It is not that the best quarterbacks in 2013 had their skills deteriorate in 2014 while the worst quarterbacks had their skills improve. It is that the best performing quarterbacks had good luck in 2013 while the worst performing quarterbacks had bad luck. The top-five quarterbacks in 2013 went from an average of 111 in 2013 to an average of 92 in 2014. The bottom-five quarterbacks in 2013 went from an average of 69 in 2013 to an average of 85 in 2014.

Figure 1
QBRs for Quarterbacks with at Least 100 pass Attempts in 2013 and 2014

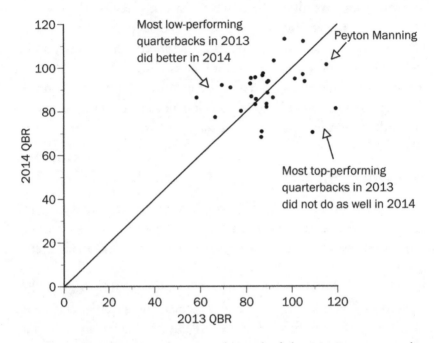

Now let's imagine that it is the end of the 2013 season and we are trying to predict how well these thirty quarterbacks will do in 2014, perhaps because we manage a football team and are thinking about trading to get a new quarterback.

A naive prediction is that players will perform the same in 2014 as in 2013. In order to take the anticipated regression into account, we can use Kelley's equation to estimate player abilities, which can then be used to predict 2014 performance. Reliability is the correlation of quarterback ratings between seasons. For predicting 2014 performance, we use the correlation between quarterback ratings for the 2012 and 2013 seasons, which is 0.43.

In comparison to the naive prediction that players will do the same in 2014 as in 2013, Kelley's equation shrinks performances toward the mean. It turns out that the shrunken forecasts are more accurate than the naive predictions by a margin of 19 to 11.

Baseball Regression

A sure sign of spring is baseball fans assembling fantasy rosters and predicting who will win the World Series. It is just as certain that fans overrate the players and teams that did exceptionally well the previous season and underrate the players and teams that did poorly.

A player who gets a base hit one out of four times at bat has a batting average of .250 and can play in the Major Leagues. Increase that to one out of three and the batting average is .333 and the player is on his way to the Hall of Fame. Baseball is said to be a game of dealing with failure in that even the best players make twice as many outs as hits.

There is skill in baseball, but there is also luck. Trying to guess what type of pitch will be thrown, deciding in less than half a second whether to swing at the pitch, using a slender rounded bat to hit a small round ball traveling 90+ miles per hour and veering in different directions, hoping that a hit ball won't go directly to a fielder.

Good players can go 0 for 4 one game and 4 for 4 the next, bat .320 one season and .280 the next. Exceptional success in a game, a month, or a season typically involves good fortune—which means that the exceptional success exaggerates the player's ability. Good fortune cannot be counted on indefinitely, so great successes are followed typically by not-so-great performances. Not necessarily bad performances, just less exceptional.

Table 1 shows the ten Major League Baseball players with the highest batting averages in 2014. Jose Altuve had the highest average, .341. Do you think that Altuve is a .400 hitter who had a disappointing season in 2014, or a .300 hitter who had a great year? Nine of the top-ten batters in 2014 did better than their career average that year. (Miguel Cabrera is the exception.) Overall, these top-ten batters had career BAs that average .298, but they averaged .322 in 2014.

There is no 2013 batting average for Jose Abreu because 2014 was his first season in the major leagues. He was selected American League Rookie of the Year in 2014, and then succumbed to the so-called "rookie-of-the-year jinx" in 2015. Except we would call it the mediocrity magnet powered by regression to the mean.

In only four of nineteen cases did a player do better in 2013 or 2015 than he did in 2014. In fifteen of nineteen cases (regression is a tendency, not a certainty), the batting averages of the top ten batters in 2014 regressed toward the mean in 2013 and 2015.

There is regression whether we look forward to 2015 or backward to 2013 because regression is a statistical phenomenon. Jose Altuve's ability did not go up in 2014 and back down in 2015. He did not focus more in 2014 or peak at the right time. Regression is not caused by fluctuations in ability, but by performances fluctuating about ability. Altuve is a really good player, with a career batting average slightly above .300, but he does not bat .300 in every game, every week, every month, or every season. Sometimes, he is lucky and bats above .300; other times, he is unlucky and bats below .300. A year like 2014, where he batted .341 and had the highest batting average in the major leagues, was a lucky year. His performance was above his ability.

Table 1
The Ten Players with the Highest Batting Averages in 2014

	2013	2014	2015	Career
Jose Altuve	0.283	0.341	0.313	0.305
Victor Martinez	0.301	0.335	0.245	0.302
Michael Brantley	0.284	0.327	0.310	0.292
Adrian Beltre	0.315	0.324	0.287	0.285
Justin Morneau	0.259	0.319	0.310	0.282
Jose Abreu		0.317	0.290	0.303
Josh Harrison	0.250	0.315	0.287	0.284
Robinson Cano	0.314	0.314	0.287	0.307
Andrew McCutchen	0.317	0.314	0.292	0.298
Miguel Cabrera	0.348	0.313	0.338	0.321
AVERAGE	0.297	0.322	0.296	0.298

Baseball pitchers also regress to the mean. One measure of a pitcher's success is his earned run average, the number of earned runs (runs not due to defensive errors) per nine innings pitched. The lower the earned run average, the better.

Table 2 shows the ten Major League Baseball pitchers with the lowest earned run averages in 2014. In only one of twenty cases did a pitcher do better in 2013 or 2015 than he did in 2014. The exception is Adam Wainwright who pitched only 28 innings in 2015 because he tore his left Achilles tendon while batting. Overall, these top-ten pitchers have career ERAs that average 3.21, but they averaged 2.31 in 2014.

As with the top-ten batters, the abilities of these pitchers did not jump dramatically in 2014 and fall back just as dramatically in 2015. It would be more apt to say that, on average, they were luckier in 2014 than they were in 2013 or 2015.

Table 2
The Ten Pitchers with the Lowest Earned Run Averages in 2014

	2013	*2014*	*2015*	*Career*
Clayton Kershaw	1.83	1.77	2.13	2.42
Felix Hernandez	3.04	2.14	3.53	3.11
Chris Sale	3.07	2.17	3.41	2.92
Johnny Cueto	2.82	2.25	3.44	3.29
Adam Wainwright	2.94	2.38	1.61	2.99
Doug Fister	3.67	2.41	4.19	3.42
Corey Kluber	3.85	2.44	3.49	3.41
Jon Lester	3.75	2.46	3.34	3.55
Cole Hamels	3.60	2.46	3.65	3.31
Garrett Richards	4.16	2.61	3.65	3.67
AVERAGE	3.27	2.31	3.26	3.21

There is nothing special about 2014. This regression is true year after year. Going back to the beginning of major league baseball, among all major league players with batting averages above .300 in any season, 80 percent did worse the previous season and 80 percent did worse the following season. Among all pitchers with earned run averages below 3.00 in any season, 80 percent did worse the previous season and 80 percent did worse the following season.

Teams also regress to the mean. Of those major league teams that win more than 100 out of 162 baseball games in a season, 90 percent did not do as well the previous season and 90 percent do not do as well the next season.

In 2014 the Los Angeles Angels had the best regular-season record, winning 60.5% of their games. The season before, they won 48.1%; the season after, 52.5%. Need I say luck? Nor did the Angels win the 2014 World Series but, then again, the regular-season champion usually doesn't win the World Series. Need I say luck? Table 3 shows that, overall, the top five teams in 2014 won 58.5% of their games that year, but only 54.1% in 2013 and 54.4% in 2015.

At the other end of the standings, the Arizona Diamondbacks won only 39.5% of their 2014 games, but they won 50.0% in 2013 and 48.8% in 2015. Table 4 shows that the bottom five teams won 41.9% in 2014, but 45.0% in 2013 and 49.6% in 2015. The mediocrity magnet in action!

Table 3
The Top Five Major League Baseball Teams in 2014

	2013	2014	2015
Los Angeles Angels	0.481	0.605	0.525
Baltimore Orioles	0.525	0.593	0.500
Washington Nationals	0.531	0.593	0.512
Los Angeles Dodgers	0.568	0.580	0.568
St. Louis Cardinals	0.599	0.556	0.617
AVERAGE	0.541	0.585	0.544

Table 4
The Bottom Five Teams in 2014

	2013	2014	2015
Houston Astros	0.315	0.432	0.531
Minnesota Twins	0.407	0.432	0.512
Texas Rangers	0.558	0.414	0.543
Colorado Rockies	0.457	0.407	0.420
Arizona Diamondbacks	0.500	0.395	0.488
AVERAGE	0.450	0.419	0.496

Table 5 shows that if we look at the average winning percentage for the top-five teams and the average losing percentage for the bottom-five teams, the numbers are pretty similar.

Table 5
The Best and the Worst Patterns Are Similar

	2013	2014	2015
Top-Five Teams, Winning Percentage	54.1	58.5	54.4
Bottom-Five Teams, Losing Percentage	55.0	58.1	50.4

Bill James, the father of the sabermetrics revolution in baseball, observed this regression year after year and called it the Law of Competitive Balance; teams with winning records tend to do not as well the next year, while teams with losing records improve. A variation on this theme, based on the same reasoning, is the Plexiglass Principle: teams that improve one year tend to fall back the next year, and vice versa.

It would be a fallacy to conclude that the skills of players and teams gyrate wildly season to season. A more reasonable explanation for the observed fluctuations in performances is that the most successful performers in any particular season generally aren't as skillful as their lofty records suggest. Most had more good luck than bad, causing one season's performance to be better than the season before and better than the season after—when their place at the top is taken by others.

The hard part is recognizing that, in all sports, player and team performances fluctuate about ability. Those who performed exceptionally well compared to others most likely also performed better than they have in the past or will in the future. The bump in their performance is most likely temporary, and their subsequent return to normal should not be a surprise.

The Paradox of Luck and Skill

One paradoxical thing about athletic performances is that at the highest levels, where all the competitors are highly skilled, the winner is likely to be determined by luck. At lower levels, where there is a wide range of skills, the winner is most likely determined by skill. For example, in golf's four major championships (Masters, U.S. Open, British Open, and PGA Championship), any one of a dozen or more players might win. Who does win is largely matter of luck. That's why there are so few repeat winners. A total of 213 golfers have won at least one of the four major golf championships; 132 (62%) of these winners won only once. Only one golfer, Bobby Jones, has won all four majors in one year. Two golfers (Ben Hogan and Tiger Woods) have won three in one year. Jack Nicklaus, the most successful golfer ever, won 18 majors over a 24-year span. It's different when four friends go out for a round of golf and the same person wins every time.

10

Jinxes, Slumps, and Superstitions

ATHLETES AND COACHES ARE TOO QUICK TO ASSUME THAT A PLAYER who has a great season will do great every season. In his Major League Baseball rookie year with Montreal Expos in 1990, second baseman Delino DeShields batted .289 and finished second in the voting for Rookie of the Year. He only batted .238 the next season, but that drop-off was generally shrugged off as the "sophomore slump" experienced by most players who have outstanding rookie seasons. The next two years, DeShields batted .292 and .295. Even though he had averaged .277 over his first four years, the Los Angeles Dodgers concluded that he was a .300 hitter who had the usual sophomore slump in 1991. The Dodgers traded a promising young pitcher for DeShields, expecting to get a player who would bat .300 for many, many years.

DeShields averaged only .241 during three seasons with the Dodgers before they traded him. DeShields played for three teams over the next six seasons, averaging .275 before he retired. Over his thirteen-year career, he averaged .268, pretty close to what he averaged his first four seasons. The Dodgers had paid too much attention to his .292 and .295 seasons and not enough attention to his .238 season. They underestimated the role of luck in the fluctuations of his batting average around .277 his first four years.

The young pitcher the Dodgers gave up to get DeShields was Pedro Martinez, who went on to win two Cy Young Awards (for the best pitcher) over an 18-year career that earned him a spot in the baseball Hall of Fame. This is widely considered the worst trade ever made by the Dodgers.

While we're on the subject, how would you explain the sophomore slump in baseball or any other sport? Regression to the mean perhaps? The sophomore slump is just a variation of the rookie-of-the-year jinx. Any athlete who is one of the top players in his or her sport during their rookie season most likely had more good luck than bad. How many athletes could perform below their ability and still have a great year?

So it is with most awards, including the Cy Young award, which is said to cause the Cy Young Jinx. One tabulation of 70 Cy Young winners found that only three did better the year after they won the award. Thirty did about the same and 37 did worse. The sportswriter who made this tabulation speculated that the pitchers wore themselves out during their Cy Young year, even though baseball players have six months to rest between seasons.

Ditto with a hall-of-fame sportswriter's report that of those baseball players who hit more than 20 home runs in the first half of the season, 90 percent hit fewer than 20 home runs during the second half. The writer concluded that there is a "second-half power outage." Perhaps because they got worn out hitting so many home runs? Or maybe they got nervous about the possibility of breaking a home run record? The regression-to-the-mean explanation is that their remarkable performance during the first half of the season involved a good deal of luck that exaggerated their skills.

Superstitions

Many professionals and amateurs are misled by regression into believing in all sorts of silly superstitions. A golfer—pro or duffer—who is performing below his or her personal average does better after switching clubs, shoes, or shirts. A baseball player who is performing below average does better after changing bats, hats, or socks. Or, even worse, an athlete who is performing below average does better when he doesn't wash his socks. Unfortunately, the regression principle predicts that players who are performing below average will usually improve, thereby confirming the value of silly superstitions.

Using a wooden stick to hit a baseball traveling 90 miles an hour and moving left, right, up, or down may be the most difficult challenge in any sport. If the ball is hit, it might go straight to a fielder for an out or it might land safely for a base hit. On average, professional baseball players only get a base hit in one out of four times at bat. From both the batter and pitcher's viewpoint, small differences separate all-stars, journeymen, and failures. Perhaps this is why baseball players are notoriously superstitious, looking for something—no matter how ridiculous—that might tilt the odds in their favor.

Wade Boggs had a terrific career playing third base, mostly with the Boston Red Sox. He had a career batting average of 0.328, played in 12 consecutive All-Star games, and was elected to the Baseball Hall of Fame in 2005. If there was a Superstition Hall of Fame, he would be in that too. Boggs woke up at exactly the same time every day and ate chicken at 2 p.m. using a 13-recipe rotation over every 14 days (he ate lemon chicken twice). For a night game at Fenway Park, the Chicken Man got to his locker at exactly 3:30, put on his uniform, and went to the dugout to warmup at 4:00. He then went through a precise warmup routine, including fielding exactly 150 ground balls. At the end of his fielding warmup, he stepped on third base, second base, first base, and the baseline (during the game he entered the field by jumping over the baseline) and taking two steps to the coach's box and four steps to the dugout. By the end of the season, Boggs' footsteps had left permanent footprints in the grass. Boggs always took batting practice at 5:17 and ran wind sprints at 7:17 (an opposing manager once tried to disorient Boggs by having the stadium clock skip from 7:16 to 7:18).

During the game, when Boggs took his position at third base, he smoothed the dirt in front of him with his left foot, tapped his glove three times, and straightened his hat. Every time he batted, he drew the Hebrew word Chai ("life") in the batter's box (though he isn't Jewish).

The source of such manic superstitions (and smaller, less disruptive ones, too) is an underestimation of the role of luck in our lives. If something good or bad happens, we assume there must be a reason. If nothing is obvious, we make something up. I changed my socks. I didn't change my socks. I touched the foul line when I ran onto the field. I didn't touch the foul line when I ran onto the field. It hardly matters. Silly superstitions are more comforting than admitting that we are at the mercy of chance.

Keep Your Mouth Shut

Here is another ludicrous example of how people are fooled by luck and regression to the mean. A pitcher is credited with a no-hitter when he pitches nine innings and the opposing team does not get any base hits, although they may get men on base via a walk, a batter being hit by a pitch, or a fielding error. A perfect game is when the pitcher retires all 27 batters with no one getting on base.

No-hitters are rare, and perfect games rarer still. As of 2015, there have been, on average about two no-hitters a season, and only 23 perfect games ever. No-hitters and perfect games are exceptional and consequently involve a lot of luck. No one throws a no-hitter while pitching below his ability.

Because no-hitters and perfect games are exceptional, a pitcher will often have a no-hitter going through the first five, six, or seven innings and then give up a hit, which ends the chance for a no-hitter. Somewhere, sometime, somebody must have mentioned to a pitcher that he had a no-hitter going. Then the pitcher happened to give up a hit and the big-mouth is blamed for jinxing the no-hitter.

In baseball, most potential no-hitters end up not being no-hitters, simply because the odds are against it. Suppose that I flip a bent coin that has a two-thirds chance of landing heads and a one-third chance of landing tails. I happen to flip heads eighteen times in a row (like pitching six innings without giving up a hit). The probability that I will flip heads nine more consecutive times (like three more innings without a hit) is less than three percent. In baseball, the chances are probably even lower because pitchers tire as the game goes on.

If, after the first eighteen heads, I happened to say, "Hey, eighteen heads in a row!," and then flip a tail sometime during the next nine tosses, would you say that I jinxed myself or would you say that it wasn't surprising because the odds were against me? The latter, I hope.

Yet, in baseball there are three unwritten rules regarding no-hitters: Don't say the word "no-hitter." Don't say the word "no-hitter." Don't say the word "no-hitter." Most players go even further, by not talking to the pitcher at all. Many players won't even sit next to the pitcher on the bench for fear that they might say something and jinx the no-hitter.

One wonderful exception was New York Yankees pitcher Don Larson, who started games two and five of the 1956 World Series against the Brooklyn Dodgers. In game two, he didn't last two innings, giving up four walks and four runs. Game five was different, very different. Through seven innings, Larson pitched a perfect game. His teammates stopped talking to him or sitting next to him, for fear they would jinx this miracle. Larson thought it was a bunch of bull, so he put down a cigarette he was smoking in the dugout and cornered a teammate, Mickey Mantle: "Look at the scoreboard, Mick. Wouldn't it be something? Two more innings to go." Well, what do you know, Larson pitched two more perfect innings, giving him the only perfect game ever pitched in the World Series.

The best newspaper headline the next day was, "The imperfect man pitched the perfect game." Larson had an otherwise forgettable career: fourteen years with eight different teams, a lifetime earned run average of 3.78, and a career win-loss record of 81-91. He was successful far beyond his ability through the first seven innings of that historic game and it would have been no surprise if he had given up a hit in the last two innings—no matter what anyone said or did. But his luck held for two more innings and he entered the record books.

It is not just baseball no-hitters and it is not just teammates. People have somehow gotten the idea that a phenomenal performance will be jinxed if a television or radio broadcaster mentions the phenomenal performance.

In November 2014, Twitter was ablaze with complaints that the Cleveland Browns had lost a football game because of the CBS announcers. Cleveland was inside their opponent's 20-yard line when the announcers told viewers that in the previous 99 times when Cleveland had gotten this close to the opponents' goal line, the team had never fumbled. This time, the Browns fumbled and the fans blamed it on CBS.

A few weeks later, as Dallas Cowboy kicker Dan Bailey was about to attempt a 41-yard field goal, the Fox announcer, Joe Buck, said that Bailey was the most accurate kicker in the history of the NFL, having made 114 of 127 field goal attempts (90 percent) in his four-year career. Wouldn't you know it, Bailey missed this one and Cowboy fans blamed Buck instead of Bailey.

How could Buck's words matter more than Bailey's leg? The broadcaster jinx is even more nonsensical than the teammate jinx. We could argue that talking to a player might make the player nervous. But we can't seriously think that a player is affected by something the player doesn't even hear. Yet some people believe this.

Even more bizarrely, I've seen people leave the room when they are watching televised sporting events for fear they will jinx their favorite athlete or team. On more than one occasion, I've seen a guy walk into a room where a football game is being televised and be delighted to discover that his favorite team is winning a game that they were predicted to lose. A few minutes later, his team fumbles the ball and he rushes out of the room, convinced that he has jinxed his team.

It's that darn selective recall playing tricks with his mind. When he watches the finish of a game and nothing special happens, he thinks nothing of it. But when there is an unexpected, unfavorable turn of events, it sticks in his mind and he thinks the coincidence is meaningful. He can't accept the fact that his favorite team might have been lucky and their luck ran out.

Remember the Good Times

Something similar happened on a travel baseball team that one of my sons played on. The coach thought very highly of himself and would often bring in a pinch hitter or relief pitcher at a crucial juncture in the game, figuring that the substitute would do better than the player he was replacing. If the player did well, the coach beamed and boasted that he was a genius. If the player did poorly, the coach blamed the kid. He couldn't accept the fact that there is luck in every player substitution, no matter how clever or random the substitution.

By remembering when his tinkering worked and forgetting when it didn't, this coach was creating his own biased version of the law of small numbers. An honest assessment would have looked at all of substitutions and tabulated the number of successes and failures, which is itself problematic because there is no way of knowing how the player who was replaced would have performed if he had been left in the game.

It's not just this coach. We all have selective memories. I still remember vividly my most glorious sports moment. I had taken up soccer at age 35 because I wanted to coach my kids' soccer teams and figured I should know something about the game I was going to coach. I captained a team in a local Sunday recreational league. We never practiced. We just showed up a half-hour before game time, laced up our cleats, and played our version of soccer.

Our team was pretty good and, one year, our final game of the season was against the British Bulldogs, captained by a salty Brit. Our teams were both undefeated and the winner of this game would get T-shirts commemorating the league championship. We had a great defense and were holding on to a 1 to 0 lead in the waning moments. Then the Bulldogs captain dribbled the ball out of his normal midfield position towards our goal. He made a convincing dive, pretending that he had been tripped, and the referee rewarded him with a penalty shot that could tie the game. By the time the referee had things sorted out, time had expired. So, the penalty shot would be the last play of the game.

Our normal goalie was absent, and I was playing goalie that day—a position I was especially ill-prepared to play. I had made few okay plays during the game, charging out of the goal to clear balls, but I had never defended against a penalty shot. I huddled with my teammates and they told me to choose a side, left or right, and dive that way.

As their star striker (ironically, a French guy playing on an otherwise all-English team) prepared to shoot, I decided to dive to my left. His shot was just inside the left goalpost, but my hand was there and knocked the ball away. The referee blew his whistle and I was mobbed by my teammates. Yay, we won T-shirts!

Decades later, I still remember that play vividly, even though it was little more than a lucky guess. There was a mild bit of athleticism in my being able to dive without hurting myself but, seriously, my decision to go left was just a mental coin flip.

What if the game had ended 0 to 0 and the two captains had agreed that instead of playing an overtime period, we would flip a coin to determine the winner? I might have called tails and won the flip and the championship, but would I remember that coin flip the rest of my life? Of course not.

I had no way of knowing that the French guy was going to shoot to my left. Half the time, my guess would have been correct; half the time, it would have been wrong. This time, I happened to guess correctly, and I will never forget it.

Is It Really Luck?

If we had nothing better to do, we could get entangled in a philosophical debate about the meaning of luck. When you flip a coin, is the outcome really luck? If we knew the precise weight and size of the coin; its exact initial resting position on our thumb and finger; the location, angle, and force with which we flick the coin with our thumb; the wind and other atmospheric conditions, couldn't we write down equations that describe the trajectory of the coin, the number of rotations, and the face-up side when it lands? In theory, we could. So, where's the luck?

Fair enough. However, in practice, we don't know enough to predict coin flips with anything other than fifty-percent accuracy. So, as a practical matter, it is useful to say that a flipped coin is equally likely to land heads or tails.

I once had a public debate at Yale with a famous theoretical statistician about the meaning of chance and probability. He argued that if he flipped a coin and caught it in his closed hand, the probability that it is heads is either zero or one, depending on how it landed. I argued that until we look at the coin it is useful to say the probability is one-half. In the same way, suppose that a deck of playing cards is shuffled and dealt for a game of poker. One of the players has four spades and wants the next card to be a spade. A classical statistician would argue that the probability is either zero or one, because once the cards have been shuffled, the order of the cards is determined. However, poker players find it helpful to use probabilities that are neither zero or one.

One last example, a woman has a suspicious lump in her breast and has a mammogram X-ray that is suggestive but not conclusive. Some statisticians argue that the probability that the lump is malignant is either zero or one, depending on whether or not it is malignant. I believe it is useful to assign a probability based on the test results so that the doctor and patient can make a decision on how to proceed.

I use the words "probability" and "luck" when something is not known with certainty. We do not know how the coin landed, whether the next card will be a spade, or whether the lump is malignant. It is useful to assign probabilities and to say that luck is involved when the coin turns out to be heads, the next card is a spade, or the lump is benign. A quarterback's success is not determined by flipping a coin, but it is useful to say that a quarterback's performance fluctuates around his ability. Otherwise, how would we anticipate correctly the regression in his performance—or the pervasive regression in education, medicine, business, and more?

11

Being Smart About It

I T IS TEMPTING TO THINK THAT PROFESSIONAL ATHLETES ARE RELENTLESSLY consistent—that some baseball players bat .250 year after year while others bat .300 every year. Nope. Suppose a player has 500 times at bat in a season. Let's make the unrealistic assumption that this player has a 30 percent chance of getting a hit every time he bats, so that he would be expected to bat .300 in the very long run. Five hundred at bats is not a very long run, statistically speaking, and his batting average could be substantially higher or lower than .300 in any one season. In fact, there is a 22 percent chance that his season average will either be below .275 or over .325.

The actual odds are even higher that he will bat outside the range .275 to .325 because the probability of getting a base hit is not a rock-solid .300. It depends on the pitcher, the ballpark, whether it is a day game or night game, the player's health, and more.

For a sense of how much player batting averages vary from year to year, Figure 1 shows major league batting averages in 2014 and 2015 for the 55 players who had at least 502 at bats in each year, the minimum number of bats required to qualify for season batting awards. Remember, these are not Little League players or overweight dads. These are the 55 major league players, professionals at the pinnacle of their profession, who got the most at bats in 2014 and 2015.

The correlation between 2014 and 2015 batting averages is 0.47, which means that there is a positive statistical relationship, but it is not a tight relationship. For any selected 2014 batting average, there is considerable variation in how the players did in

2015. The only consistency is that batting averages regress toward the mean.

Points above the 45-degree line are batters who did better in 2015 than in 2014; points below the line are batters who did worse in 2015 than in 2014. The batters with the highest averages in 2014 tended to do not so well in 2015, while the batters with the lowest batting averages in 2014 tended to do better in 2015. On average, a player who batted 50 points above (or below) average in 2014 batted only about 23 points above (or below) average in 2015.

Figure 1
Major League Batting Averages in 2014 and 2015

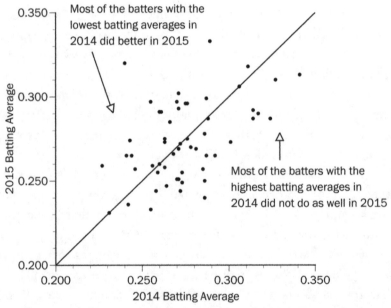

There is nothing special about 2014 and 2015. Every single year, there is only a loose relationship between how players perform that season and how they perform the next season. I looked at 115 years of baseball data, from 1901 through 2015, and found that the average correlation between batting averages in adjacent years was 0.50. Some players have more ability than other players, but there is a lot of luck in how any player performs in any given year.

Figure 2
Major League Earned Run Averages in 2014 and 2015

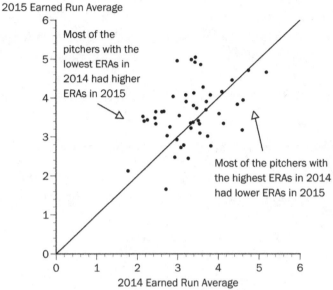

Figure 2 shows comparable data for pitchers' earned run averages in 2014 and 2015 for the 50 players who pitched at least 162 innings in each year, the minimum number of innings to qualify for pitching awards. The correlation is 0.39 and, again, the mediocrity magnet pulls the best and worst performers towards the mean.

On average, pitchers with an earned run average one run below (or above) average in 2014 had earned run averages only 0.40 runs below (or above) average in 2015. For the 115-year time period as a whole, the average correlation between earned run averages in adjacent years is 0.31, substantially lower than the year-to-year correlation in batting averages.

Better Predictions

Now, suppose that we are trying to predict how well a major league baseball player will perform next season based on how well he did this season. A naive prediction would be that a player who batted .300 this year will bat .300 next year. We can do better because we know that luck causes batting averages to vary substantially from season to season. Anticipating regression toward the mean, we want to take into account the likelihood that those players who performed far above their peers probably benefited from good luck and are not as far from their peers in ability as their performance suggests.

We can use Kelley's equation to estimate ability from performances. Here, our prediction of a player's 2015 batting average (BA) is his estimated ability, based on his 2014 batting average, the overall batting average for all players in 2014, and the reliability of batting averages:

predicted 2015 BA = R(2014 BA) + (1 − R)(average 2014 BA)

The reliability can be estimated from the correlation between batting averages for the 2013 and 2014 seasons. Since the reliability is less than one, we predict each player's performance by shrinking his 2014 batting average towards the overall average batting average in 2014.

I did exactly that using Major League Baseball data for the years 1901 through 2015. This was not a phony prediction based on data that were not, in fact, available at the time of the prediction. My 2015 predictions were based on 2013 and 2014 data; my 2014 predictions were based on 2012 and 2013 data; and so on.

For the most restrictive calculations, I only used data for players who had at least 502 times at bat. The shrunken predictions based on regression toward the mean were more accurate 59 percent of the time. Using a broader data base had little effect on the results. Looking at players with at least 400 times at bat, the shrunken forecasts were still more accurate 59 percent of the time. Looking at players with at least 300 times at bat, the shrunken forecasts were again more accurate 59 percent of the time.

The results were similar for pitchers. Oddly enough, for players who pitched at least 162 innings, the shrunken forecasts were again more accurate 59 percent of the time. Looking at players with at least 130 innings pitched, the shrunken forecasts were once again more accurate 59 percent of the time. Looking at players with at least 100 innings pitched, the shrunken forecasts were more accurate 60 percent of the time.

Beating the Spread

People who bet on football games say that, "It doesn't matter whether a team wins or loses, but whether it beats the spread." The point spread set by bookmakers ("bookies") is a prediction of the margin of victory. If, for example, Minnesota is a 4-point favorite to beat Chicago, people who pick Minnesota win their bets if Minnesota wins by more than four points; those who pick Chicago win their bets if Chicago wins or if Minnesota wins by fewer than four points.

Typically, for every $11 wagered, the payoff for a winning bet is $10. Bookies are not bettors. Their goal is to set the point spread to equalize the number of dollars wagered on each team. If they can do this, the losers pay $11, of which winners get $10 and the book-maker $1, no matter how the game turns out.

A bettor wagering $11 against $10 has to win 52.38 percent of the time to break even. An expert who can win more than 52.38 percent can make money betting on football games. There are so many people who think they are experts, and so few who actually are, that bookies can make a good living taking bets from the experts and the delusional.

Bookies try to take into account all the factors that bettors take into account: player strengths and weaknesses, injuries, home-field advantage, historic rivalries. If bookmakers are aware of systematic bettor irrationalities, they will take these into account, too. For example, if gamblers are inclined to bet on the Green Bay Packers for sentimental reasons, bookmakers will set the point spread as if Green Bay is a better team than it really is. If, objectively, the expected value of the Packers' margin of victory is three points, bookmakers might make the Packers a 4-point favorite in order to equalize the dollars wagered for and against the Packers. The "smart money" will bet against the Packers and win more than half the time. Fortunately for bookies, there is more dumb money than smart money.

Experienced bookmakers are evidently good at setting betting lines. Otherwise, they wouldn't stay in business. Of course, gambling is a zero-sum game in that if bookies make money, it must be because bettors lose money. Despite their losses, gamblers keep betting because of a mistaken faith in their ability to pick winners. When they win, it is because they are smart. When they lose, it is because of bad breaks and bad officiating. It's not their fault.

Bettors are correct when they say that the outcome depends on chance—unpredictable variations in play calling, injuries, officiating, and even the proverbial bounces of the football. They are incorrect, however, in their belief that luck only affects the bets they lose. Their winning bets may be lucky, too.

The real lesson to be drawn from the importance of luck in football games is that the differences in team performances we see exaggerate the underlying differences in abilities that we are trying to gauge. Performances consequently regress toward the mean, in that those teams that perform the best one week typically do not perform as well the next week.

If gamblers understood regression, the betting lines set by bookmakers would take regression into account since they want an equal number of dollars wagered on the opposing teams. If, on the other hand, gamblers do not appreciate regression sufficiently, they will overreact to successes and failures, and cause betting lines to do likewise.

Marcus Lee (then a student, now a doctor) and I investigated this question. Our betting strategy was based on the simple presumption that a blindness to regression causes gamblers to overrate successful teams and to underrate unsuccessful ones. When a team beats the point spread, one possible explanation is that the team's ability is better than previously believed; we should consequently revise our estimate of the team's ability upward. Another factor is luck. The regression argument suggests that the amount by which a team beats the spread overstates its ability. If bettors do not fully appreciate this, they will think too highly of the team and push the point spread too high the next week (this team will be too big a favorite). If so, it may be profitable to bet against teams that beat the point spread and to bet on teams that did not.

In each game, our strategy was to bet against whichever of the two opposing teams has been doing better against its point spreads. We did not estimate the teams' abilities. We simply assumed that the bettors push the spread too high for teams that have been the most successful in beating the spread.

Our initial hypothesis was that bettors are influenced by the number of points by which a team beats the spread, and that we should consequently base our bets on the cumulative points by which teams are over or under the spread. As an example, consider the game between Minnesota and Miami in the second week of the 2000 season. The previous week, Minnesota had been a four-point favorite against Chicago and won by three points. Because Minnesota is now one point under the spread for the 2000 season, we are inclined to bet $1 on Minnesota, hoping that its failure to beat the spread will make other betters, on average, too pessimistic about its chances against Miami. In its first-week game, Miami had been a 3-point favorite over Seattle, and won by 23 points. Because Miami is 20 points over the spread, we want to bet $20 against Miami. On balance, wanting to bet $1 on Minnesota and $20 against Miami, we bet $21 on Minnesota. As it turned out, Minnesota was a 3-point favorite and won by six points, giving us a win.

An alternative strategy is to count the number of times a team has beaten or failed to beat its spreads. In our example, going into the second week of the season, Minnesota had failed to beat its spread once and Miami had beaten its spread once; so, we bet $2 on Minnesota.

Each week, we scaled the bets so that the total amount wagered was $1,000, but this scaling maintained the structure of our bets: betting the most on games where the difference in the past performance of the two teams was the largest. If two teams had done about the same against the spread, we bet very little. If one team had done much better against the spread than its opponent, we placed a relatively large bet on the opponent.

A referee for the journal that published our paper suggested a slight modification. Our strategy is based on our theory that gamblers underestimate the role of chance. Logically, this myopia will be most exploitable when there is a large difference between how the opposing teams performed in previous weeks. The referee suggested that, instead of betting small amounts when the two opposing teams have done about the same, don't bet anything. We looked at cutoffs ranging from zero to ten using this rule: don't bet on a game unless the difference in points or games between the two opposing teams is greater than the cutoff. For instance, if one team has beaten the spread five times and failed twice (a net of three) and the other team has beaten the spread once and failed six times (a net of minus five), the difference between the two teams is eight. If we are using a cutoff of eight or lower, we bet on the team doing poorly. If our cutoff is more than eight, we wait for a more alluring situation.

Table 1 provides a summary of our data. The favorite covered the spread if it won by more than the point spread; otherwise, the favorite failed to cover the spread. The point spreads were, on average, remarkably accurate, with the favorite covering the spread 912 times and failing to do so 914 times. So a simple strategy of always betting on the favorite (or always betting on the underdog) would have failed. The average point spread and the average actual margin of victory shown in Table 1 are from the standpoint of the team favored to win the game. Thus, over the time period we studied, the average point spread was 5.60 points and the average margin of victory for the favored team was 4.98 points.

The correlation between the point spread and the actual margin of victory was 0.40, which means there was substantial regression to the mean.

Table 1
Point Spreads and Outcomes

	Number of Games	Favorite		Average Point Spread	Average Victory	Correlation
		Covered	Failed			
1993	224	102	113	6.07	4.75	0.39
1994	224	100	116	5.20	3.94	0.41
1995	240	114	120	5.73	4.62	0.38
1996	240	128	106	5.48	5.28	0.39
1997	240	102	120	5.42	4.52	0.35
1998	240	127	100	5.73	6.87	0.49
1999	248	124	114	5.51	4.21	0.31
2000	248	115	125	5.85	5.57	0.46
1993-2000	1,904	912	914	5.60	4.98	0.40

Table 2
Profits

	Wagers Based on Cumulative Points			Wagers Based on Cumulative Games		
Points Cutoff	Number of Bets	Percent Won	Percent Profit	Number of Bets	Percent Won	Percent Profit
0	1,733	52.05	3.73	1,445	52.87	5.96
2	1,671	52.18	3.74	1,199	52.79	6.09
4	1,614	52.17	3.79	563	53.46	9.63
6	1,560	52.18	3.71	228	57.02	22.47
8	1,502	52.60	4.15	79	58.23	31.91
10	1,426	52.31	4.15	19	73.68	152.53

Table 2 shows a summary of the results of wagers based on the two strategies:

1. The cumulative points a team has been over or under its spreads
2. The cumulative games that a team has been over or under its spreads

For example, using a cutoff of zero (no cutoff), we won 52.05 percent of the wagers based on the cumulative number of points that the opposing teams were over or under the spread. Wagering $11 against the bookie's $10, the net profit was equal to 3.73 percent of the total dollars wagered. For wagers with no cutoff based on the cumulative number of games that a team has been over or under its spreads, our strategy won 52.87 percent of the wagers, with a 5.96 percent net profit. (Our strategy was sometimes profitable, even though we did not win 52.38 percent of our bets, because we did not wager the same amount on every game.)

The use of cutoffs generally increased the winning percentage and the profit—supporting the idea that our strategy works because bettors are excessively influenced by unexpectedly strong or weak performances.

Our interest was purely academic—to provide convincing empirical evidence that gamblers do not fully appreciate the important role of luck in athletic successes. However, our research has reportedly been used profitably by an Australian betting on rugby games and by an American and a Canadian betting on NFL games.

VI. HEALTH

12

Take Two Aspirin

DURING A ROUTINE PHYSICAL CHECKUP SEVERAL YEARS AGO, I GOT the usual height and weight measurements, a few questions about my lifestyle (no, I don't smoke), and a battery of tests. The nurse measured my temperature, heart rate, and blood pressure. Urine and blood samples were taken and tested for who knows what. That evening, I got a telephone call telling me that one test (I don't remember which one) came back worrisome. Ninety-five percent of all healthy patients have readings in the "normal range." One of my test results was outside the normal range, so I was evidently not healthy.

My doctor said, "Not to worry." She told me to take two aspirin, get a good night's sleep, and come back the next day to be retested. I did what I was told and, to my relief, my next reading was back inside the normal range. Was it the aspirin or the good night's sleep? Probably neither. Most likely, it was just another case of regression to the mean.

There are almost always variations in test results for perfectly healthy individuals. Blood pressure depends on the time of day, digestion, and our emotional state. Cholesterol can be affected by what we've eaten or whether we exercised before the test. Every test is prone to equipment errors and to human errors in reading, recording, or interpreting the results.

Suppose that I was completely healthy and that, on every test, I had a 95 percent chance of getting a reading inside the normal range. This means that every test has a 5 percent chance of a reading outside the normal range. If I take ten independent tests, there is a 40 percent chance that at least one test will, by luck alone, be judged abnormal.

If a test result is abnormally high or low because of chance fluctuations, a second test will probably yield a result closer to the mean. This regression makes it difficult to assess a patient's true condition. It also makes it very hard to assess the value of medical treatments—in my case, whether the aspirin and good night's sleep had any effect at all.

The consequences are more serious than being worried unnecessarily. Abnormal readings due to chance fluctuations can lead to unnecessary treatments. The subsequent regression to the mean in the test results can lead to an unfounded belief that the treatment was effective. My doctor might have prescribed an expensive medication with possibly dire side effects. When my readings returned to normal, we both would have concluded erroneously that the medication worked.

It is not just doctor-prescribed medications. The same is true of individuals who try home remedies. People who feel sick, tired, or achy are likely to try something different. Even if "something different" is worthless, they will, on average, experience improvements—which then become testimonials for the effectiveness of "something different."

It is also true of doctors and even hospitals. Remember the student pilots who were praised for doing well and screamed at for doing poorly? The high performers tended to do not as well on their next flight, while the low performers tended to do better than before, but not because of the praise or screams. In the same way, well-performing hospitals that are rewarded tend to have a dip in performance, while poorly-performing hospitals that are punished tend to improve. The evident—but mistaken—conclusion is that the way to improve hospital performance is to take away the rewards and increase the punishments.

Better, Worse, Better

An article in the *Journal of the American Medical Association* showed how regression can lead to erroneous conclusions when medical treatments are monitored over time. At every point in the monitoring, there is random noise in the readings because of measurement errors and physical changes in the patient that are unrelated to the treatment. Because of this random noise, it is commonplace to find that some patients initially show improvement while others appear to be adversely affected. Then, as the monitoring proceeds, there is a puzzling reversal. The patients who showed the most improvement seem to relapse, while those who were adversely affected initially now benefit from the treatment.

The assessments are an imperfect measure of the treatment's effects. So, those patients with the largest measured improvements tend, on average, not to have benefitted as much as the measurements suggest. Later assessments will be closer to the mean, indicating that the initial benefits have been reversed. At the other end of the spectrum, the initial adverse effects that are observed may be noise. Later assessments will be more positive, suggesting that the negative effects were reversed.

The authors illustrated this argument with data from a two-year study of the effects of alendronate on the hip bone mineral density (BMD) of middle-aged women who were at risk of osteoporosis. Overall, these women gained an average of 2.2 percent BMD during the first year and an additional 0.9 percent during the second year. However, misleading reversals appear when the data were disaggregated.

Figure 1 shows the BMD gains during the first and second years, when the women are separated into eight groups based on the improvement during the first year. The top group gained an average of 10.4 percent the first year, but lost an average of 1.0 percent the second year. The bottom group lost an average of 6.6 percent the first year and gained an average of 4.8 percent the second year.

It is easy to draw the wrong conclusions. For the patients who show the most improvement initially and then seem to fare worse, the doctors might conclude that the treatment only works in the short run and needs to be terminated before it causes more harm than good. For those patients who seemed to be adversely affected initially, the doctors might stop the treatment and then see the patients improve—seemingly confirming the conclusion that the treatment was bad for them. In both cases they would be fooled by regression toward the mean.

Figure 1
Average Bone Mineral Density Gains during the First and Second Years

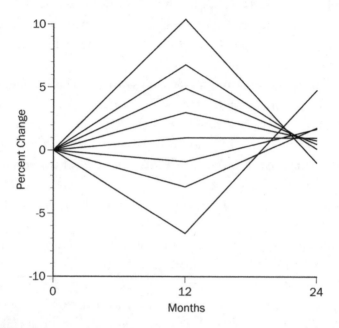

Self-Healing

It has been said that, "If properly treated, a cold will be gone in fourteen days; left untreated, it will last two weeks." This is part of the fundamental wisdom of the age-old advice offered by doctors who sound like they are trying to avoid being bothered when they say, "Call me in the morning."

Even if I had been ill and the aspirin had done nothing at all, I might have felt better in the morning because of the body's remarkable ability to heal itself. Think of a scrape severe enough for you to bleed. The body clots the blood, forms a scab, and repairs the skin—all by itself, with no medical intervention.

These are two distinct reasons why "call me in the morning" works. First, medical tests are imperfect measures of a patient's condition. If the test results are worrisome, the patient's condition is most likely not as bad as the initial results suggest, so the results regress. Second, people who actually are ill often improve as their bodies fight off what ails them. So, their condition regresses towards normal.

Post Hoc Reasoning

The fact that "call me in the morning" often works means that any treatment that is prescribed will be given credit for the improvement, even if the treatment is worthless. An old cartoon shows a dog barking, then a volcano erupting, then the citizens bowing down to the dog-god that makes volcanos erupt. This sort of reasoning is a logical fallacy known as *post hoc ergo propter hoc* ("after this, therefore because of this"). The fact that one event happened shortly after another doesn't mean that the first event caused it to happen.

Suppose, for example, that the measurement of a patient's blood pressure is unusually high, suggesting an elevated risk of disease. A medication or change in behavior (more exercise, less junk food) is recommended and—voila!—the measurements improve. However, we can expect the measurements to improve, on average, even if the patient does not take a medication or change his or her behavior. The elevated blood pressure may have been a measurement error or reflect natural variation. Perhaps the patient rushed through the parking lot to avoid being late for an appointment. Perhaps the patient was nervous about the tests.

The same is true when a group of people are studied. Suppose that physical exams are given to a large number of people and the five percent with the highest cholesterol readings are identified and given a special diet. Remember how the students with the lowest test scores can expect their scores to improve if a tutor does nothing more than wave a hand over their heads? In the same way, we can expect the cholesterol readings to improve even if the dietary instructions are nothing more than, "Wave a hand over your food before eating it."

Here is an example that combines test taking and drugs. In 1987, the *New York Times* reported an Associated Press story headlined "Drug May Help the Overanxious on S.A.T.'s." Supported by a grant from the American Academy of Pediatrics, Dr. Harris Faigel, the Director of University Health Services at Brandeis University, identified 25 high school students who took the SAT at the end of their junior year in high school and had done worse than expected based on their IQ test scores and other academic metrics.

An hour before retaking the SAT in their senior year, these underperforming students were given propranolol to relax them. Their senior-year test scores improved, on average, by 50 points on the verbal part of the SAT and by 70 points on the mathematics part, compared to typical improvements of 18 and 20 points for students who retake the SAT.

Dr. Faigel offered a theory about why propranolol worked: "Their parents and teachers had convinced them that, if they didn't do well on the SATs, they'd never get into college. The result was they approached the SATs with a tremendous amount of anxiety and fear." However, regression to the mean is an obvious explanation for their improvement. Just like the low-scoring students given special tutoring, the abilities of these 25 students were no doubt higher than their scores. Indeed, they were selected precisely because of evidence that their abilities were higher than their SAT scores indicated. So, we can expect their scores to improve whether or not they take propranolol.

The unfortunate consequence of this *New York Times* article was that some parents read the article and had their children take propranolol before taking the SAT. I say "unfortunate" because, as Dr. Faigel observed, one consequence of taking propranolol is that students may become drowsy—not exactly what one would want when taking the SAT.

There are other, more bizarre, treatments that, like propranolol, are based on little more than fallacious *post hoc* reasoning. Somebody isn't feeling well, endures a treatment, and improves. It must be the treatment that caused the improvement, right?

Believe it or don't, some people practice "urine therapy," rubbing their urine on their skin or drinking it in order to cure allergies, asthma, and other ailments. One of several Internet web sites promoting urine therapy claims that, "Multiple sclerosis, colitis, lupus, rheumatoid arthritis, cancer, hepatitis, hyperactivity, pancreatic insufficiency, psoriasis, eczema, diabetes, herpes, mononucleosis, adrenal failure, allergies and so many other ailments have been relieved through use of this therapy." I think I'll pass.

Going back in time, for centuries otherwise well-informed people believed that epidemics were caused by breathing foul-smelling air (especially night air). When the stench was unbearable, people stayed indoors or, if they went out, covered their noses and mouths with a cloth. Some thought that bad outdoor air could be combatted by creating equally foul indoor air; for example, keeping a goat inside the house or breathing flatulence that had been thoughtfully stored in jars.

Even more farfetched was the "Powder of Sympathy" advocated by Sir Kenelm Digby, a prominent seventeenth century Englishman who was honored for his naval victory over the Dutch and Venetians at Scanderoon and was a founding member of the Royal Society, established to support and promote science. Digby wrote a book that went through 29 editions recommending this salve for wounds: "take Roman vitriol [copper sulphate] six or eight ounces, beat it very small in a mortar, shift it through a fine sieve when the sun enters Leo; keep it in the heat of the sun and dry by night." The oddest thing about this potion is not that it must be made when the sun enters Leo but that, instead of applying the salve to the wound, it is applied to the weapon that caused the wound. A person who had been cut by a knife was advised to rub the powder on the knife. No doubt, wounds sometimes healed, but it equally certain that the healing was because of the body's recuperative power—not because of the Powder of Sympathy.

In our own time, one of my sons tried a treatment that (to my relief) is more benign than rubbing urine or breathing flatulence. He is a baseball player and, as noted earlier, baseball is said to be a game of failure in that even the best hitters make twice as many outs as hits. Many players deal with this frustration by trying almost anything— not washing their socks, not shaving, stepping on the foul lines when entering the field and jumping over foul lines when exiting.

A few years ago, many players—professional and amateur— became convinced that they would be better hitters if they wore braided titanium necklaces. One company advertised that, "Materials emit energy that is effective in controlling the flow of bio-electric current in ones body. Improves the alignment of ions when this current is stabilized (so called 'Minus Ion Power'), especially at the body's crucial motor joints." Baseball necklaces are also said to improve a player's concentration and focus and speed recovery from sports fatigue. Like many fads, this one seems to be passing. My son no longer wears a baseball necklace. A necklace from the company whose ad I quoted had a list price of $39.99, but was now advertised for $2.73 with free shipping.

First, Do No Harm

The cost of believing in treatments that do not work is not just the cost of the treatment, but the possible side effects. In addition to drowsiness, some possible (admittedly rare) side effects of propranolol include fever, rash, vomiting, diarrhea, impotence, and heart failure. The decision to take a drug or have some other medical treatment should be based on a comparison of the prospective benefits and risks. That comparison is thrown off if regression to the mean leads us to overestimate the benefits.

Another reason why the benefits may be exaggerated is that medical trials focus on patients who are known to have a certain illness. Outside the trials and inside the doctors' offices, treatments are prescribed for patients who are not as ill as the people who were tested. In some cases, patients have the symptoms, but not the illness. The benefits are consequently smaller, but the harm is the same—which tilts the true benefit/harm balance towards harm.

For example, antibiotics are widely viewed as a miracle drug and they are often very effective. However, some doctors seem to prescribe antibiotics reflexively even when they are more likely to do harm than good. The possible side effects include allergic reactions, vomiting, or diarrhea. The risks for patients who are given antibiotics are the same regardless of whether there are any benefits. The unfortunate consequence of over-prescribing antibiotics is that, too often, the risks are real and the benefits are fictional.

For childhood ear infections, the American Academy of Pediatrics now recommends that, instead of prescribing antibiotics, parents and doctors wait and watch to see if the body can fight off the infection unaided. More generally, *The ICU Book*, the best-selling and widely respected guidebook for intensive care units, advises: "The first rule of antibiotics is try not to use them, and the second rule is try not to use too many of them."

Observational Data

In a 1973 letter to the prestigious *New England Journal of Medicine*, a doctor named Sanders Frank reported that 20 of his male patients had diagonal creases in their earlobes and also had many of the risk factors (such as high cholesterol levels, high blood pressure, and heavy cigarette usage) associated with heart disease. For instance, the average cholesterol level for his patients with noticeable earlobe creases was 257 (mg per 100 ml), compared to an average of 215 for healthy middle-aged men.

Before you rush to look in a mirror, one very big problem with this report is that these are observational data, not any sort of clinical trial. Anecdotes aren't evidence and data is not the plural of anecdote.

Dr. Frank observed earlobe creases in 20 of his patients and also noticed that they have elevated risk factors. So? Most people don't like going to the doctor unless they think they have a serious health problem. Dr. Frank specialized in respiratory medicine, and high cholesterol, blood pressure, and cigarette usage are all associated with respiratory problems. Perhaps these patients came to Dr. Frank because cholesterol, blood pressure, and cigarettes were affecting their breathing. Any trait they happen to share (balding, beady eyes, big thumbs) will seem to explain these elevated risk factors, even if they are completely unrelated.

Subsequent studies investigating the relationship between earlobe creases and heart disease have been inconclusive, perhaps because of the ambiguities in defining earlobe creases. The most reasonable explanation may be that as people age, they tend to develop earlobe creases and are also increasingly at risk for heart disease, but neither necessarily causes the other. Instead of poking around their ears, we can ask their age.

Over and over, statisticians say that correlation is not causation, yet many are unpersuaded. If we see a correlation, we assume there must be a reason. Maybe so, maybe not. If A is correlated with B, it may be that A is causing B, that B is causing A, or that something else is causing both A and B. Or maybe it is just a fluke, like the coincidental correlation between the number of people who drowned by falling into a swimming pool and the number of films Nicolas Cage appeared in. There is a web site "Spurious Correlations" that has this one and hundreds of other hilarious flukey correlations.

No one takes the correlations at this web site seriously (at least, I hope not), but other correlations are more subtle and do get taken seriously. Data from six large medical studies found that people with low cholesterol levels were more likely to die of colon cancer; however, a later study concluded that the low cholesterol levels may have been caused by colon cancer that was in its early stages and therefore had not yet been detected. It wasn't that A caused B, it was that B caused A.

For centuries, residents of New Hebrides believed that body lice made a person healthy. This folk wisdom was based on the observation that healthy people often had lice and unhealthy people usually did not. However, it was not the absence of lice that made people unhealthy, but the fact that unhealthy people often had fevers, which drove the lice away.

Cancer Clusters

In the 1970s, Nancy Wertheimer, an epidemiologist, and Ed Leeper, a physicist, drove through Denver, looking at homes that had been lived in by people who had died of cancer before the age of 19. They tried to find something—anything—these homes had in common. They found that many cancer-victim homes were near large power lines, so they concluded that exposure to the electromagnetic fields (EMFs) from power lines causes cancer.

A journalist named Paul Brodeur wrote three *New Yorker* articles that reported other anecdotal correlations between power lines and cancer. He ominously warned that, "Thousands of unsuspecting children and adults will be stricken with cancer, and many of them will die unnecessarily early deaths, as a result of their exposure to power-line magnetic fields."

The resulting national hysteria offered lucrative opportunities for consultants, researchers, lawyers, and gadgets, including Gauss meters that would allow people to measure EMFs in their own homes (rooms with high EMF readings were to be blocked off and used only for storage). Fortunately, the government did not tear down the nation's power lines.

The problem with this scare is that, even if cancer is randomly distributed among the population, there will, more likely than not, be a geographic cluster of victims. To demonstrate this, I created a fictitious city with 10,000 residents living in homes evenly spaced throughout the city, each having a one-in-a-hundred chance of cancer. (I ignored the fact that people live in families and that cancer is related to age.) I used computerized coin flips to determine the cancer victims in this imaginary town. The resulting cancer map is shown in Figure 2. Each black dot represents a home with a cancer victim. There are no cancer victims in the white spaces.

Figure 2
A Cancer Map

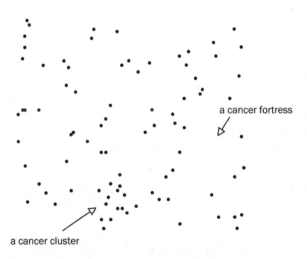

a cancer fortress

a cancer cluster

There is clearly a cancer cluster in the lower part of the map. If this were a real city, we could drive through the neighborhood where these people live and surely find something special. Perhaps the city's Little League fields are nearby. If we now compare cancer rates for people who live near these Little League fields with cancer rates for people who live far away, guess what? The cancer rates are higher near the fields, suggesting that living near a Little League field causes cancer.

Figure 2 also shows a cancer fortress, a part of town where nobody has cancer. If we drive through this cancer-free neighborhood, we are sure to find something unusual. Perhaps the town's water tower is nearby. If we now compare the cancer rates for people who live near the water tower with cancer rates for people who live far away, cancer rates are of course lower near the water tower. That's why we chose that neighborhood. Because nobody there has cancer.

In each case, near the Little League fields and near the water tower, we have the same problem. If we use the data to invent the theory (Little League fields cause cancer, water towers protect against cancer), then of course the data support the theory! How could it be otherwise? Would we make up a theory that did not fit the data? Of course not.

A theory cannot be tested fairly by looking at the data that were used to invent the theory. We need fresh data. Chance correlations are an extreme example of regression to the mean in that observations that are all luck will vanish when tested with new data.

Other studies in other countries found no relationship between EMFs and cancer. Experimental studies of rodents found that EMFs far larger than those generated by power lines had no effect on mortality, cancer incidence, immune systems, fertility, or birth abnormalities.

Weighing the theoretical arguments and empirical evidence, the National Academy of Sciences concluded that power lines are not a public health danger and there is no need to fund further research, let alone tear down the power lines. One of the nation's top medical journals weighed in, agreeing that we should stop wasting research resources on this issue.

In 1999, the *New Yorker* published an article titled "The Cancer-Cluster Myth," which implicitly repudiated the earlier articles written by Paul Brodeur. Nonetheless, the idea that cancer clusters are meaningful lives on. The Internet has government-sponsored interactive maps that show the incidence of various cancers by geographic area, all the way down to census blocks. Millions of dollars are spent each year to maintain these maps with statistics that are up to date, but potentially misleading. One interactive site has cancer mortality rates for twenty-two types of cancer, two sexes, four age groups, five races, and more than three thousand counties. With millions of possible correlations, some are bound to be frightening.

To accommodate this fear, the U.S. Center for Disease Control and Prevention (CDC) has a web page where people can report cancer clusters they discover. Even though the CDC cautions that, "Follow-up investigations can be done, but can take years to complete and the results are generally inconclusive (i.e., usually, no cause is found)," more than a thousand cancer clusters are reported and investigated each year.

The Placebo Effect

People who believe in the power of medicine may be especially responsive to treatment, even if the treatment has no medical value. In some cases, patients who are given worthless treatments show re-markable physical improvement. In other cases, the outcomes are vague—my headache is gone, my back feels better—and it is not clear if the patients are just saying what they think they are supposed to say. In many cases, it is hard to separate the placebo effect from the expected regression caused by fluctuating measurements and the body's ability to self-heal.

For example, our knee joints can become trashed with tears, strains, and bits of loose cartilage after decades of supporting our overweight bodies while we stand, walk, run, jump, and dance. The most common treatment, used hundreds of thousands of times each year, is arthroscopic surgery. Two small incisions are made, one for a small fiberoptic camera and one for the miniature instruments used for the surgery. The surgeon removes the debris and then repairs, cleans, smooths, and trims what is left. Five thousand dollars later, the pain is gone. At least that's the theory.

For decades, arthroscopic surgery was not compared to any-thing. The doctors did the surgery, the patients said they felt better, so what more do we need to know? Well, for one thing, were their knees really improved? Maybe it was just a placebo effect. Maybe pa-tients feel better because they are hopeful that the procedure works and confident that the doctors know what they are doing. Or maybe patients say they feel better because they know that is what they are supposed to say.

The way to tell if surgery makes a real difference is to do a controlled experiment in which randomly selected patients have arthroscopic surgery while others do not. A seemingly insurmountable difficulty is that the patients would know whether they had surgery, and this might influence how they felt and their reports about how they felt.

To get around this problem, an elaborate ruse was used for an experiment conducted in the 1990s using 180 military veterans. For those in the comparison group, two superficial incisions were made and the doctors acted out a charade that mimicked arthroscopic surgery. The patients did not know that they were part of an experiment. Nor did the doctors who independently assessed the veterans' conditions over the next two years. The study concluded that at no point did those patients who had real arthroscopic surgery feel less pain or function any better than the comparison group that had the phony surgery.

After this study was published in the *New England Journal of Medicine* in 2002, another study (published six years later in the same journal) confirmed that patients with knee osteoarthritis who received arthroscopic knee surgery, medication, and physical therapy fared no better on measures of pain, stiffness, and physical functions than did a comparison group that only received medication and physical therapy. Many doctors now advise patients to forgo the surgery.

Something very similar happened with gastric freezing—a now-discredited remedy for stomach ulcers. Stomach ulcers can be excruciatingly painful and were once commonly treated by surgically cutting off the supply of acid to the stomach. A doctor thinking outside the box thought that the pain might be reduced by using ice to numb the stomach, in much the same way that ice is used to reduce the pain from a sprained ankle. However, swallowing dozens of ice cubes would be unpleasant and inefficient, since there is no way of ensuring that the ice cubes make sustained contact with the ulcer.

The goofy solution was to insert a balloon into the stomach of an ulcer sufferer and pump a supercooled fluid through this balloon. This would be less expensive and less dangerous than surgery, though the effects would not be as long-lasting. Experiments done in the 1950s concluded that this wacky idea actually worked, in that patients reported reduced acid secretions and stomach pains. After the results were published in the *Journal of the American Medical Association*, gastric freezing was used for several years to ease the suffering of ulcer patients.

Like arthroscopic surgery for knee problems, there was no comparison group and, thus, no way of knowing if it really worked. There may have been a placebo effect or, when asked about their stomach pains, the patients may have been inclined to give an optimistic answer. What was needed was an experiment in which randomly selected patients had supercooled fluid pumped into them while another group received a body-temperature fluid. Of course, none of the patients could be told which fluid they were receiving.

When this experiment was eventually done, the results were surprising. While 34 percent of those receiving the gastric-freezing treatment reported improvement, 38 percent of those receiving the body-temperature fluid did too! Evidently, the reported relief from ulcer pain was due entirely to the placebo effect. Subsequent studies confirmed that gastric freezing has no beneficial effects and doctors stopped using it.

Control Groups

Taken together, measurement errors, the body's natural recuperative power, the power of positive thinking, and cheerful reporting mean that we can't tell if a purported medical treatment works unless we compare what happens to what would have happened if the person hadn't taken the treatment.

A valid study should compare the group receiving the treatment to a control group that does not receive the treatment. In medical research, the control group is usually given a placebo, which appears identical to the treatment but has no medical value. In studies of vitamin C, those in the control group are given pills that look and taste like vitamin C, but are in fact made of inactive substances that do not affect the body. In the arthroscopic knee surgery experiments, those in the control group were given look-alike surgery. In the gastric freezing experiments, those in the control group were given dummy balloons.

In another example, medical researchers once tested a vaccine for the common cold using 548 college students who felt they were especially susceptible to colds. Half were given the experimental vaccine and the other half, the control group, were given plain water. The group taking the vaccine reported an astonishing 73 percent decline in colds in comparison with the previous year—astonishing, except for the fact that the control group reported a 63 percent decline! Apparently, this particular season was relatively cold-free, or there were psychological benefits from the presumed vaccine, or the students reported a decline in colds because they thought that this is what the researchers wanted to hear.

If there had been no control group, the researchers would have concluded that the vaccine was a miraculous success. Having a control group for comparison, they concluded that the vaccine had little, if any, effect. The importance of a control group is made even more striking by the researchers' report that they received several letters from doctors similar to this one: "I have a patient who took your cold vaccine and got such splendid results that he wants to continue it. Will you be good enough to tell me what vaccine you are using?" When they checked the records, the researchers often found that the patient with such splendid results had received plain water.

It is also important that the patients be randomly assigned to the treatment group and the control group. If we simply compare people who choose to take vitamin C with people who choose not to, our results might be biased by the fact that there are systematic differences between the people who make these choices. Perhaps one group is more concerned about their health and is therefore more likely to exercise regularly and have a healthy diet. It is these choices, not the vitamin C, that make a difference.

An example of this bias occurred in tests of whether an experimental vaccine called BCG could reduce deaths from tuberculosis. A group of doctors gave the vaccine to children from some tubercular families and not to others, apparently depending on whether parental consent could be obtained easily. Over the subsequent six-year period, 3.3 percent of the unvaccinated children and 0.67 percent of the vaccinated children died from tuberculosis. The vaccine apparently reduced the risk of death by 80 percent.

However, a second study was then conducted in which doctors restricted their study to children who had parental consent. Half were given the vaccine, half were not. This time, there was no difference between the death rates for vaccinated and unvaccinated children. The initial study was evidently biased by systematic differences between the parental-consent and non-consent groups. Perhaps the non-consent families had lifestyles that made fatal tuberculosis more likely. We don't know the reason, but we do know that randomization is needed so that the results are not affected by the choices people make.

Another example was a 1971 study that found that people with bladder cancer were more likely to be coffee drinkers than were people who did not have bladder cancer—suggesting that coffee causes bladder cancer. However, a confounding factor was that people who chose to drink coffee were also more likely to choose to smoke cigarettes. Was it the coffee or the cigarettes that was responsible for the bladder cancer?

Doctors could not do an experiment in which randomly selected people were forced to drink coffee, while others were not allowed to. However, they could use statistical procedures that accounted for the confounding effects of smoking. In essence, they looked at people who smoked the same number of cigarettes, but drank different amounts of coffee, and people who drank the same amount of coffee, but smoked different numbers of cigarettes.

In 1993, a rigorous analysis of 35 studies placed the blame on cigarettes and exonerated coffee. For a given level of coffee consumption, smokers were more likely to get bladder cancer. For a given level of smoking, coffee drinkers were not more likely to get bladder cancer. The study concluded that there is "no evidence of an increase in risk of [lower urinary track cancer] in men or women after adjustment for the effects of cigarette smoking." A 2001 study confirmed that tobacco increases the risk of bladder cancer and coffee does not, but added a new twist: smokers who drank coffee were less likely to develop bladder cancer than were smokers who did not drink coffee. Coffee seemed to partly offset the ill effects of tobacco.

Double Blind

Researchers, too, can be affected by knowing that an experiment is being conducted and what results are expected. It is natural to hope that a new treatment has the desired effects. To encourage honest reporting by both subjects and researchers, well-designed studies are double-blind: neither the subject nor the researcher knows who is in the treatment group and who is in the control group until all the data have been collected.

A famous experiment conducted in a senior-level course in experimental psychology is a good example of how researchers can swayed by their expectations. Each of 12 students was given five rats to test in a maze. Six of the students were told that their rats were "maze-bright" because they had been bred from rats that did very well in maze tests; the other six students were told that their rats were "maze-dull." In fact, there had been no special breeding. The students were given randomly selected ordinary rats. Nonetheless, when the students tested the rats in ten runs each day for five days, the rats that had been labeled maze-bright had higher scores each day and their scores also increased more over time. The results seemed to indicate that the bright rats were learning faster than the dull rats, but really just demonstrated that the researchers were fooling themselves.

The examples in this chapter illustrate why an insufficient appreciation of the role of luck in our lives can lead us to see illness when we are not sick and to see cures when treatments are worthless. To alleviate this confusion, the gold standard for medical tests is a double-blind experiment with the subjects randomly assigned to the treatment and control groups. Yet, even then, as the next chapter explains, regression to the mean can fool us.

13

The Tin Standard

TO DETERMINE WHETHER A MEDICAL TREATMENT REALLY WORKS, the gold standard is a randomized controlled trial:

1. In addition to the group receiving the treatment, a control group receives a placebo, so that we can compare treatment to no treatment without worrying about the placebo effect or the body's natural ability to heal.

2. The subjects are randomly assigned to the treatment group and the control group, so that we don't need to worry about whether the people who choose the treatment are systematically different from those who don't.

3. The test is double-blind so that the subjects and researchers are not influenced by knowledge of who is getting the treatment and who is not.

When the study is finished, the statisticians move in. The statistical issue is the probability that, by chance alone, the difference between the two groups would be as large as that actually observed. Most researchers consider a probability less than 0.05 to be "statistically significant." Differences between the treatment and control groups are considered statistically persuasive if they have less than a 1-in-20 chance of occurring by luck alone.

Unfortunately, even with all these safeguards, medical tests are far from infallible.

We all know that some pain-relievers work better for some people and other pills work better for others. That's true of most medical treatments. They are neither 100 percent effective or 100 percent ineffective. When the effects are modest and vary from patient to patient, the results of a medical test depend on which persons are randomly assigned to the treatment group, which are put in the control group, and which are omitted from the study.

Suppose that, if left untreated, a certain illness would show noticeable improvement due to self-healing in ten percent of the people with this illness. If a worthless treatment is tested, it might turn out that, by luck alone, the treatment group includes a disproportionate number of the ten percent of the population who self-heal, while the control group has a disproportionate number of the 90 percent who do not. If so, this worthless treatment may seem to work miracles.

Now consider a test of an effective treatment that shows noticeable improvement in 20 percent of the people with the illness. Just by luck, the treatment group might include a disproportionate number of the 80 percent of the population who won't benefit from the treatment, while the control group has a disproportionate number of people who self-heal. If so, the treatment seems to backfire in that the people given the treatment are less likely to show improvement.

The larger the sample, the less likely it is that the luck of the draw will mislead us. With two flips, a coin may land tails both times. With 1,000 flips, the fraction that land tails is very likely to be close to fifty percent. Unfortunately, many medical studies are small and even large studies can be misleading.

A standard level of statistical significance is five percent. This means that if the treatment tested is worthless, there is only a five percent chance it will show such positive benefits. Fair enough, but this also means that five percent of all worthless treatments tested will show statistically significant results.

Researchers chasing fame and funding can generate statistically significant results simply by testing lots of treatments. Even if they are so misguided as to test nothing but worthless remedies, they can expect five out of every hundred worthless treatments tested to turn out to be statistically significant—which is enough to generate published papers and approved grant requests.

Pharmaceutical companies can make enormous profits from treatments that are clinically "proven" to be effective. One way to ensure that some treatments—it hardly matters which ones—will be endorsed is to test thousands of treatments. No matter how many statistical hurdles are required, chance alone ensures that some worthless treatments will clear them all.

There is no incentive for a company to retest a treatment that has been approved and is generating millions of dollars in revenue, and little incentive for independent researchers to do so either. Where is the payoff in confirming that something works or, for that matter, raising doubts?

A worthless treatment that appears to be effective is a false positive. There are also false negatives, cases in which an effective treatment does not show statistical significance. Consider a test with a five percent chance of a false positive—a five percent chance that a test of a worthless treatment will find a statistically significant difference between the treatment group and the control group. For simplicity, assume a simple ten percent chance of a false negative—a ten percent chance that an effective treatment will not show a statistically significant effect.

If there is only a five percent chance of a false positive and a ten percent chance of a false negative, it seems we should be able to tell the difference between worthwhile and worthless almost every time. Not necessarily. It depends on how many tested treatments really are effective. Table 1 shows the implications if ten percent of all tested treatments are effective and ninety percent are useless.

Of every 10,000 treatments tested, 100 are effective; 90 of these 100 treatments will show statistically significant effects and 10 will be false negatives. Of the 9,900 worthless treatments tested, 495 will be statistically significant (false positives). Overall, there will be 585 studies with statistically significant test results, but only 90 are really effective. An astonishing 85 percent of all treatments "proven" to be effective are actually worthless.

This paradox is related to a common confusion about conditional probabilities. One hundred percent of the players in the English

Premier League are male, but only a fraction of one percent of all males play in the Premier League. Here, 90 percent of all effective treatments are statistically significant, but only 15 percent of all statistically significant treatments are effective.

Table 1
85 Percent of All Proven Treatments are Ineffective

	Significant	Not significant	Total
Effective treatments	90	10	100
Ineffective treatments	495	9,405	9,900
Total	585	9,915	10,000

The assumption that ten percent of all tested treatments are effective is likely to be generous. Researchers are so eager to find cures that they may test 99 worthless treatments for every treatment that actually does anything. If so, calculations like those in Table 1 show that 98 percent of all treatments proven to be effective are worthless.

These kinds of calculations are the basis for a famous paper with the provocative title, "Why Most Published Research Findings Are False," written by John Ioannidis, who holds positions at the University of Ioannina in Greece, the Tufts University School of Medicine in Massachusetts, and the Stanford University School of Medicine in California.

Ioannidis has devoted his career to warning doctors and the public about naively accepting medical test results that have not been replicated convincingly. His famous paper with the scandalous title works out the math as we did, although his assumptions are more damning than ours and the dismal probabilities even bleaker.

In addition to these theoretical calculations, Ioannidis has compiled a list of real-world "proven" treatments that turned out to be ineffective. In one study, he looked at 45 of the most widely respected medical studies published during the years 1990 through 2003. In only 34 cases were attempts made to replicate the original test results with larger samples. The initial results were confirmed in 20 of these 34 cases (59 percent). For seven treatments, the benefits were much smaller than initially estimated; for the other seven treatments, there were no benefits at all. Overall, only 20 of the 45 studies have been confirmed, and these were for the most highly respected studies! The odds are surely worse for the thousands of studies published in lesser journals.

Ioannidis guesstimates that 90 percent of published medical research is flawed. Even when there are follow-up studies that discredit the initial positive results, the memorable cures are usually remembered and the later cautions overlooked, so that doctors continue to recommend bogus treatments.

The Funnel Plot

By its very nature, the results of random samples depend on the luck of the draw. Suppose that a certain medication has, on average, substantial medical benefits for ten percent of the population. In any particular random sample, the number of people who benefit may be larger or smaller than ten percent. In a sample of ten people, there is only a 39 percent chance that exactly one person will benefit. There is a 35 percent chance that nobody will benefit and a 26 percent chance that more than one person will benefit. This means that in a large number of perfectly fine ten-person studies, 35 percent of the studies will find no benefit and 26 percent of the studies will overestimate the benefit.

The chances that a study will substantially overestimate or underestimate the effect depends on the size of the sample. As the sample size goes up, the variation in the study results goes down. This is shown in Figure 1, which is known as a funnel plot. In this hypothetical case, ten percent of the population would report a substantial improvement in their condition if they were in the control group, either because of the placebo effect or their bodies' natural healing power. In contrast, twenty percent of the population would report a substantial improvement if they received the treatment. The medical treatment causes a ten-percentage-point increase in the number of people who experience substantial improvements, which is an impressive doubling of the odds.

The horizontal axis shows the estimated benefit found in any single study. The average improvement for the population is ten percent; however, individual studies may show somewhat larger or smaller benefits due to the inherent randomness of random samples.

The sample size shown on the vertical axis is the total sample. A sample of 100 is 50 people in the control group and 50 in the treatment group.

The sides of the funnel show the range in which 95 percent of the studies are expected to fall. With a sample of size 50, for example, there is a 95 percent probability that the estimated effects will be between -4% (four percent worse than the control group) and +24% (24 percent better than the control group.) The shape of the funnel shows that as the sample size increases, the variability across samples decreases, although even for samples of 1,000 people, the 95 percent range in estimated benefits is from 6 percent to 14 percent.

Figure 1
The Effect of Sample Size on the Range of Test Results

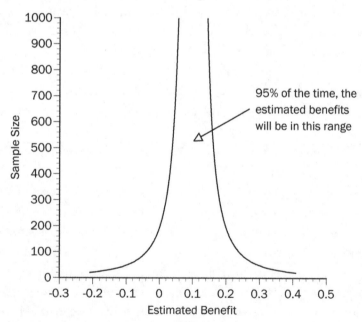

To illustrate this variability, and how the variability diminishes as the sample size increases, I simulated 100 studies with sample sizes ranging from 20 to 1,000. The dots in Figure 2 show these 100 estimates of the benefits. As expected, the estimated benefits are roughly symmetrical around ten percent, and there is more variability with smaller sample sizes. Four of the 100 studies happen to be outside the funnel, about what we expect.

Figure 2
One Hundred Simulated Studies Confirm the Funnel Pattern

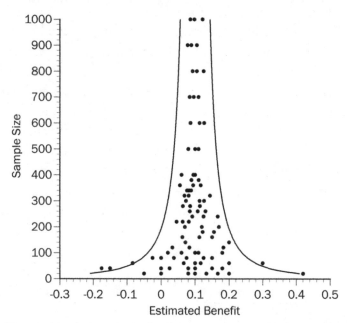

Figure 1 is theory. Let's look at practice. A 1993 article summarized the published results of ten randomized trials testing the use of intravenous magnesium to treat severe heart attacks. Overall, 6.2 percent of the patients who were given magnesium died compared to 9.9 percent of the control group. Magnesium apparently reduced the risk of death by nearly 40 percent. There was also a one-third reduction in abnormal heart rhythms (7.6 percent versus 11.3 percent) for those who survived. The authors concluded that, "These data suggest that magnesium given to patients during [severe heart attacks] can produce significant reductions in mortality and serious morbidity."

One of the co-authors of this study was Salim Yusuf, an extremely distinguished cardiologist. He is enshrined in the Canadian Medical Hall of Fame, where he is credited with publishing more than 800 articles in refereed journals and was the second most cited researcher in the world in 2011. Surely, his recommendations should be followed.

Figure 3 shows the fatality data for the ten studies that were analyzed. The horizontal axis is the difference between the survival fractions for the magnesium recipients and the control groups. For example, a six percent fatality rate versus a ten percent fatality rate is a four-percentage-point increase in the survival rate. Figure 4 shows similar data for abnormal heart rhythms.

What is striking about these graphs is that left-hand side of the funnel is missing. If we take the single large study as a reasonable approximation of the benefit (two percentage points for fatalities and one percentage point for abnormal rhythms), the smaller studies should be evenly distributed on both sides of these numbers. Yet, seven of the nine smaller studies showed greater benefits than found in the large study—in most cases, much greater benefits.

The most likely explanation is publication bias. A large number of researchers might test a treatment. Even if it is worthless, some tests will show statistical significance by luck alone. Or a single researcher might test dozens, hundreds, or even thousands of treatments and only report the ones that are statistically significant. Either way, the tests that find statistically significant benefits are likely to be published in medical journals and the tests that don't are likely to be put in the circular file known as a trash can.

Studies that find little or no benefit tend not to be submitted to journals and, if they are submitted, are likely to be rejected. This disparity in outcomes is particularly acute for small studies where there is a large variability in the results. Small studies are more likely to find promising benefits, which is what medical journals and doctors are looking for.

It is very misleading to aggregate nine small studies with one large study because the nine small studies are surely a biased measure of the results of all small studies, many of which were never published. If these "missing" studies had been included in the analysis, the overall estimated benefits surely would have been more modest.

Ironically, the review article co-authored by Dr. Yusuf concluded that, "When judged on its own, [the large study] may be regarded by some physicians as not being sufficiently persuasive to change their practice. However, when viewed in the context of the overall data, the evidence becomes much more conclusive."

Figure 3
Difference in Fatalities for Magnesium Recipients and Controls

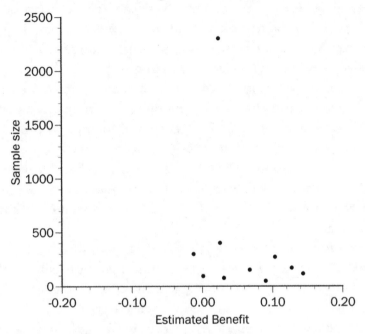

Figure 4
Difference in Abnormal Heart Rhythms for Magnesium Recipients and Controls

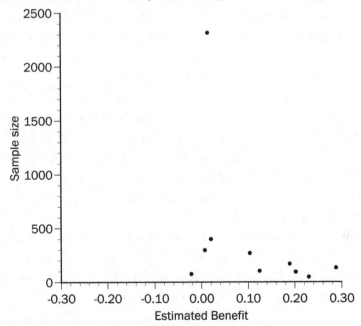

Dr. Yusuf acknowledged that the small studies are crucial for his conclusion, but he overlooked the likelihood that the small studies are misleading because of publication bias. The reality is that the large study is probably the best estimate of the benefits of magnesium for heart attack victims and these estimated benefits are small.

Several more small studies were published in the next few years, again supporting the benefits of magnesium (and again tainted by publication bias). Then another large study was reported in 2002, this time with 6,213 patients, and found that there was absolutely no benefit from magnesium, either in the entire sample or in several subgroups. The authors concluded that, "In view of the totality of the available evidence, in current coronary care practice there is no indication for the routine administration of intravenous magnesium."

An independent review of the evidence, published by the well-regarded Cochrane organization, concluded that "it is unlikely that magnesium is beneficial in reducing mortality," and noted that there are several documented risks of serious side effects.

Some doctors were puzzled. If there is, in fact, little or no benefit, how could so many small trials produce statistically significant benefits? One doctor editorialized, "The probability of that occurring by chance is extremely remote." He concluded that the discrepancy between the large studies and the small studies must have had something to do with the doses administered or the age of the patients (even though the studies took dosage and age into account).

This view has become a minority. Most well-informed researchers now accept that small studies can show benefits where there are none because studies showing benefits are more likely to be published than are studies that find no benefits. Funnel plots are now used to investigate whether unpublished studies are likely to have been overlooked. Statistical procedures have even been developed to account for the missing studies.

However, problems remain. In the magnesium studies, the publication bias was revealed by the existence of large studies that found little or no benefit. But imagine that there had been only nine small studies. As shown in Figure 5, there would be no evidence of publication bias, and the benefits would seem clear. In most medical research, there are only a few studies, often small, and no way of knowing whether the reported results are misleading because of publication bias.

Figure 5
No Sign of Publication Bias

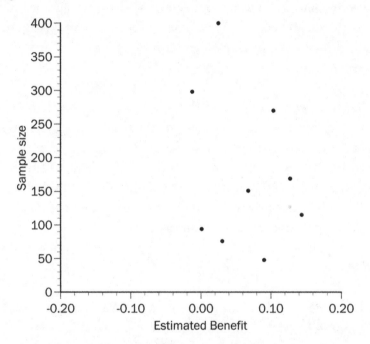

The general principle is that, because of publication bias, treatments that are reported to be beneficial tend to be less beneficial than reported. Small studies have the most variability and the highest risk of exaggerated benefits. This wouldn't be a problem if there were a large number of small studies and they were all reported. We could then average the tests to give, in effect, a large study with unbiased data. The problem is that if we only see the studies that had positive results, we overestimate the treatment's benefit. When the treatment is then applied to patients, the results regress to the mean.

Ransacking Data

Statistical tests assume that researchers start with well-defined theories and gather appropriate data to test their theories. Some people work the other way around. Test every theory you can test—whether the theories make sense or not—and concoct a rationale for whatever result turns out to be the most statistically significant.

Some researchers are not only unapologetic about this approach, they even encourage it. Daryl Bem, a prominent social psychologist wrote that,

> The conventional view of the research process is that we first derive a set of hypotheses from a theory, design and conduct a study to test these hypotheses, analyze the data to see if they were confirmed or disconfirmed, and then chronicle this sequence of events in the journal article. . . . But this is not how our enterprise actually proceeds. Psychology is more exciting than that.

He goes on:

> Examine [the data] from every angle. Analyze the sexes separately. Make up new composite indexes. If a datum suggests a new hypothesis, try to find further evidence for it elsewhere in the data. If you see dim traces of interesting patterns, try to reorganize the data to bring them into bolder relief. If there are participants you don't like, or trials, observers, or interviewers who gave you anomalous results, place them aside temporarily and see if any coherent patterns emerge. Go on a fishing expedition for something—anything—interesting.

Using the fishing-expedition approach, Bem was able to find evidence to support some truly remarkable theories. For example, in a paper titled, "Feeling the Future," Bem reported that when erotic pictures were shown in random locations on a computer screen, his subjects were able to guess beforehand, with 53 percent accuracy, whether the picture would be on the left or right side of the screen. Part of Bem's fishing expedition was that he did his experiment with five different kinds of pictures, and chose to emphasize the one that was statistically significant.

The journal that published "Feeling the Future" published another paper a year later titled "Correcting the Past," coauthored by four professors at four different universities. They reported that seven experiments attempting to replicate Bem's claim that people can feel the future "found no evidence supporting its existence."

The same problem occurs with medical research. If a treatment doesn't show statistical significance for the entire sample, see if it works for subsets. Separate the data by gender, race, age. Try different age categories. If the treatment didn't help the problem you were initially testing, see if there were other beneficial effects. The problem with significance-chasing is that one can always find patterns in data, even in randomly generated numbers, if only one looks hard enough. As Ronald Coase cynically observed, "If you torture the data long enough, it will confess."

I'll Have Another Cup

In the early 1980s, a group led by Brian MacMahon, a widely respected researcher and chair of the Harvard School of Public Health, found "a strong association between coffee consumption and pancreatic cancer." This study was published in *The New England Journal of Medicine*, one of the world's premier medical journals, and reported nationwide. The Harvard group advised that pancreatic cancer could be reduced substantially if people stopped drinking coffee. MacMahon followed his own advice. Before the study, he drank three cups a day. After the study, he stopped drinking coffee.

One problem was that MacMahon did the study because he thought there might be a link between alcohol or tobacco and pancreatic cancer. He looked at alcohol. He looked at cigarettes. He looked at cigars. He looked at pipes. When he didn't find anything, he kept looking. He tried tea. He tried coffee and finally found something: patients with pancreatic cancer drank more coffee.

Suppose that six independent tests are conducted, in each case involving something that is, in fact, unrelated to pancreatic cancer. There is a 26 percent chance that at least one of these tests will find an association that is statistically significant at the 5 percent level—a 26 percent chance of making something out of nothing.

MacMahon's study had another flaw. He compared hospitalized patients who had pancreatic cancer to patients who had been hospitalized by the same doctors for other diseases. The problem was that these doctors were often gastrointestinal specialists, and many of their patients had given up coffee because of fears that it would exacerbate their ulcers and other gastrointestinal problems. The patients with pancreatic cancer had not stopped drinking coffee. This virtually guaranteed that there were more coffee drinkers among the patients with pancreatic cancer. It wasn't that drinking coffee caused pancreatic cancer, but that other health problems caused people who did not have pancreatic cancer to stop drinking coffee.

Subsequent studies, including one by MacMahon's group, failed to confirm the initial study. This time, they concluded that, "in contrast to the earlier study, no trend in risk was observed for men or women." The American Cancer Society agreed: "the most recent scientific studies have found no relationship at all between coffee and the risk of pancreatic, breast, or any other type of cancer."

Not only has subsequent research not confirmed MacMahon's initial theory, it now appears—at least for men—that drinking coffee reduces the risk of pancreatic cancer!

Distant Healing

In the 1990s a precocious young doctor named Elisabeth Targ investigated whether patients with advanced AIDS could be healed by distant prayer. Twenty patients with advanced AIDS were randomly divided into two groups, with ten patients sent prayers and positive energy from self-proclaimed healers living an average of 1,500 miles away. The six-month study was double-blind in that neither the patients or the doctors knew which patients were being prayed for. (The healers knew who they were praying for.) Four of the 20 patients died (the expected mortality rate), but none were in the prayer group.

Encouraged by these results, Targ did a second six-month study involving 40 AIDS patients divided into two double-blind groups. Photos of those in the prayer group were sent to 40 experienced distance healers (including Buddhists, Christians, Jews, and shamans) who took turns trying to work their magic. This study found that those in the prayer group spent fewer days in the hospital and suffered from fewer AIDS-related illnesses. The results were statistically significant and published in a prestigious medical journal.

Targ became a celebrity because of her provocative findings and scientific rigor. People with agendas cited her work as proof of the existence of God or the inadequacy of conventional views of mind, body, time, and space. Targ did not speculate about why it worked. She just knew that it worked.

She was awarded a $1.5 million grant from the National Institute of Health (NIH) for an even larger study of AIDS patients and for investigating whether distant healers could shrink malignant tumors in patients with brain cancer. Shortly after being awarded this grant, Targ was discovered to have brain cancer herself and, despite being sent prayers and healing energy from all over the world, died four months later.

After her death, problems were discovered with her research. By the luck of the draw, the four oldest patients in her original study had been put in the non-prayer group. The fact that all four died may have been because of their age, not the absence of prayer. This is a good example of the fragility of small samples. With large samples, it becomes less likely that there will be important differences between the control group and the treatment group.

Targ's second study, involving 40 patients, had planned to compare the mortality of the prayer and non-prayer groups. However, one month after the six-month study began, triple-cocktail therapy became commonplace and only one of the 40 patients died—which demonstrated the effectiveness of the triple cocktail but eliminated the possibility of a statistical comparison of the prayer and non-prayer groups.

Targ and her colleague, Fred Sicher, turned instead to physical symptoms, quality of life, mood scores, and CP4+ counts. There were no differences between the prayer and non-prayer groups. Elisabeth's father had done experiments attempting to prove that people have paranormal abilities to perceive unseen objects, read each other's minds, and move objects just by willing them to move. He told his daughter to keep looking. If you believe something, evidence to the contrary is not important. Just keep poking through the data for evidence that supports your beliefs. Finally, she found something—hospital stays and doctor visits, although medical insurance was surely a confounding influence.

Then Targ and Sicher learned about a paper listing 23 AIDS-related illnesses. Maybe they could find a difference between the prayer and non-prayer groups for some of these 23 illnesses. Unfortunately, data on these illnesses had not been recorded while her double-blind test was being conducted. Undeterred, Targ and Sicher pored over the medical records of their subjects even though they now knew which patients were in the control group and which were in the prayer group. When they were done, they reported that the prayer group fared better than the non-prayer group for some illnesses.

Their published paper suggested that the study had been designed to investigate these few illnesses and did not reveal that the illness data were assembled after the study was over and the double-blind controls had been lifted. Perhaps they found what they were looking for simply because they kept looking. Perhaps they found what they were looking for because the data were no longer double-blind. One of the peer reviewers of the study said that he would have evaluated it quite differently had he known about the data snooping.

Targ's NIH study continued after her death. It found no meaningful differences in mortality, illnesses, or symptoms between the prayer and non-prayer groups. An even larger study, conducted by the Harvard Medical School, looked at 1,800 patients who were recovering from coronary artery bypass graft surgery. The patients were randomly assigned to three groups: some patients were told they would receive distant prayer and did; other patients were told they might receive prayer and did; the final group were told they might receive prayer but did not. There was no difference between the prayer and non-prayer patients but, oddly enough, those who were told they definitely would receive prayer were more likely to develop complications than were those who were only told that they might receive prayer. Perhaps there was some sort of perverse-placebo effect?

Unfortunately, a lot of published medical research is little more than random variation in the data that is significant only if we over-look the fact that these flukes were uncovered by testing lots of theories or inventing theories to match coincidental patterns in the data. They vanish in subsequent tests. There is a full and complete regression to zero effect.

This pattern is so common in medical research, it even has a name—the "decline effect." Some researchers who have seen the decline effect first-hand with their own research are so perplexed that they set off on wild-goose chases looking for a causal explanation, when the reason is right in front of them: regression. If the initial positive findings were due to luck—random variations exploited by multiple tests or tortured data—it is no surprise that the subsequent results are often disappointing.

Too Much Chaff, Not Enough Wheat

Many miraculous treatments, like insulin and smallpox vaccines, have been discovered and proven effective through medical research. The problem for society is that too many good researchers and too many valuable resources are devoted to medical studies that are fundamentally flawed by data torturing and publication bias.

Not only are resources squandered, but the credibility of medical science is undermined by an exasperating cycle of medical "never minds." Annual or biannual mammograms have long been recommended for women over the age of forty; now we are told mammograms may do more harm than good. Tens of millions of prescriptions of the antidepressant Prozac are filled each year; now we are told Prozac may be just a placebo. Coffee and chocolate used to be bad for us; now they are good for us.

Too often, "doctors practicing medicine" seems to mean that they are practicing until they get it right. Doctors keep telling us to never mind because the initial research was flawed—usually because of publication bias or because the data were ransacked to find something publishable.

VII. BUSINESS

14

The Triumph of Mediocrity

I
T IS PARTICULARLY IRONIC THAT THE ECONOMICS PROFESSION PRODUCED what is arguably the most famous regression fallacy of all time and, yet, economists—even Nobel laureates—continue to make the same error over and over again. It is a fallacy that will not die.

The story starts in the 1930s with Horace Secrist, a distinguished economics professor at Northwestern University. Secrist wrote thirteen textbooks and was Director of Northwestern's Bureau of Economic Research. In the depths of the Great Depression, he published his *tour de force*, a book titled *The Triumph of Mediocrity in Business*, based on ten years of research. His *Triumph* was 468 pages long, with 140 tables and 103 charts documenting his great effort and supporting his profound conclusion.

Secrist and his assistants spent ten years collecting and analyzing data for 73 different industries, including department stores, clothing stores, hardware stores, railroads, and banks. He compiled annual data for the years 1920 to 1930 on several metrics of business success, including the ratios of profits to sales, profits to assets, expenses to sales, expenses to assets. For each ratio, he divided the companies in an industry into quartiles based on the 1920 values: the top 25 percent, the second 25, the third 25, and the bottom 25. He then calculated the average value of the ratio for the companies in the top quartile in 1920 for every year from 1920 to 1930. He did the same for the companies in the second, third, and bottom quartiles in 1920. In almost every case, the companies in the top two quartiles in 1920 were closer to the average in 1930, and the companies in the bottom two quartiles in 1920 were also closer to the average in 1930.

He had evidently discovered a universal economic truth. American business was converging to mediocrity. Hence, he titled his book *The Triumph of Mediocrity in Business*. Secrist summarized his conclusion:

> Complete freedom to enter trade and the continuance of competition mean the perpetuation of mediocrity. New firms are recruited from the relatively "unfit". . . . Superior judgment, merchandizing sense, and honesty . . . are always at the mercy of the unscrupulous, the unwise, the misinformed, and the injudicious. The results are that retail trade is overcrowded, shops are small and inefficient, volume of business inadequate, expenses relatively high, and profits small. So long as the field of activity is freely entered, and it is; and so long as competition is "free," and, within the limits suggested above, it is; neither superiority or inferiority will tend to persist. Rather mediocrity tends to become the rule.

The nation's economic problems were apparently due to the new economic principle he had discovered: competitive pressures inevitably dilute superior talent. The evident solution? Protect superior companies from competition with less-fit companies.

Before publishing his work, Secrist asked 38 prominent statisticians and economists for comments and criticism. They apparently had no reservations. After publication, the initial reviews from eminent colleagues were unanimous in their praise.

> This book furnishes an excellent illustration of the way in which statistical research can be used to transform economic theory into economic law, to convert a qualitative into a quantitative science.—*Journal of Political Economy*

> The author concludes that the interaction of competitive forces in an interdependent business structure guarantees "the triumph of mediocrity." The approach to the problem is thoroughly scientific.   —*American Economic Review*

The results confront the business man and the economist
with an insistent and to some degree tragic problem.
              —*Annals of the American Academy of*
                  Political and Social Science

Then Harold Hotelling wrote a devastating review that politely
but firmly demonstrated that Secrist had wasted ten years proving
nothing at all. What Secrist became famous for was being fooled by
regression to the mean.

In any given year, the most successful companies are more likely
to have had good luck than bad, and to have done well not only relative to other companies, but also relative to their own "ability." The
opposite is true of the least successful companies. This is why the subsequent performance of the top and bottom companies is usually
closer to the average company. At the same time, their places at the
extremes are taken by other companies experiencing fortune or misfortune. These up-and-down fluctuations are part of the natural ebb
and flow of life and do not mean that all companies will soon be
mediocre.

It is just like our earlier data on batting averages. The best and
worst performers in any given year are more nearly average the next
year, but this does not mean that everyone will wind up with the same
mediocre batting average.

To illustrate these natural fluctuations in business, I simulated a
hypothetical industry with 100 firms and calculated profits as a rate
of return on assets. Some firms have higher profits, on average, than
other firms, but every firm's profits are sometimes higher and sometimes lower than their average profits. I then used a set of random
numbers to generate each firm's profits in one year, which I called
1920. I did the same thing for a second year, which I called 1930.

Following Secrist, I grouped the firms into quartiles based on
their 1920 profits. Table 1 shows the average profits in 1920 and,
also, the average profits in 1930 for the firms in each 1920 quartile.
For example, the 25 firms in the top quartile in 1920 had average
profits of 48 in 1920 and 39 in 1930. Their profits regress in 1930
because they were lucky in 1920.

Table 1
Average Profits, Quartiles Formed Using 1920 Profits

	1920	*1930*
First quartile, 1920	48.0	39.0
Second quartile, 1920	34.6	32.3
Third quartile, 1920	25.4	27.6
Fourth quartile 1920	12.0	21.0

Using the data in Table 1, Figure 1 is a visual demonstration of the fact that the firms in each 1920 quartile were more nearly average in 1930.

Figure 1
Quartiles Based on 1920 Profits Regress in 1930

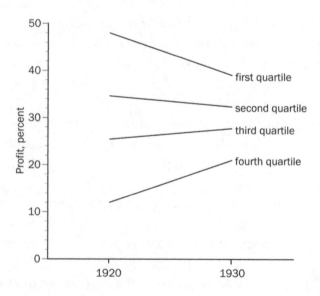

Just as abnormal parents generally have less abnormal children, and vice versa, profits regress whether we go forward or backward in time. I used the same procedure to generate profits for 1910. Table 2 shows that firms in the top two quartiles in 1920 were not only more nearly average in 1930, they were also more nearly average in 1910. So were the firms in the bottom two quartiles in 1920. Figure 2 shows this convergence before and after 1920, the year used to form quartiles.

Table 2
Average Profits, Quartiles Formed Using 1930 Profits

	1910	*1920*
First quartile, 1930	39.0	48.0
Second quartile, 1930	32.3	34.6
Third quartile, 1930	27.7	25.4
Fourth quartile, 1930	21.0	12.0

There is absolutely no convergence in abilities. I assumed that abilities are the same for each firm in each year, 1910, 1920, and 1930. In the real world, abilities do evolve over time. I assumed that abilities are constant in order to demonstrate that profits regress even if abilities never change.

The regression in 1910 and 1930 is just a reflection of the fact that profits bounce around and that we separated firms into groups that were generally lucky or unlucky in 1920. As Hotelling put it, "These diagrams really prove nothing more than that the ratios in question have a tendency to wander about." Ten years down the drain.

Figure 2
Quartiles Formed Based on 1920 Profits Regress in 1910 and 1930

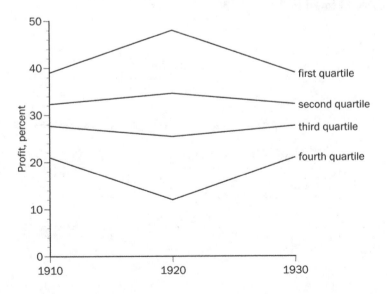

Regression toward the mean does not imply that every company will soon be equally mediocre any more than it implies that everyone will soon be the same height, have the same batting average, or get the same score on tests. Regression simply reflects the fact that when an unobserved trait is measured imperfectly, measurements that are far from the mean overstate how far the underlying trait is from the mean.

These are hypothetical data. How about real data? I calculated the profit rates of 28 companies that were in the Dow Jones Industrial Average at some point during the years 1990-2010 and had profit data for the years 1990, 2000, and 2010. Because of ups and downs in the overall economy, the average profit rates were different in these three years: 6.6 percent, 8.1 percent, and 6.2 percent. So, I calculated the difference between each firm's profits and the average profit that year. Quartiles were then formed based on the 2000 data, and the average profit difference was calculated for each of these quartiles for 1990, 2000, and 2010.

Figure 3 shows that, just as with the hypothetical data in Figure 2, profits tend to regress before and after the year used to form the quartiles.

Figure 3
Quartiles Formed in 2000 for 28 Dow Stocks

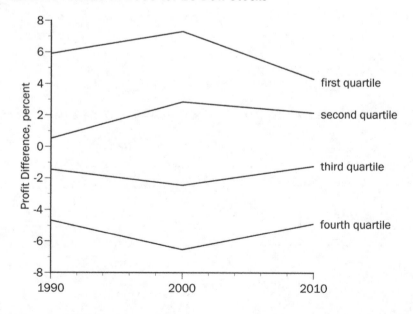

Do Old Fallacies Ever Die?

Even though Secrist's error was clearly dissected by Hotelling, the error lives on. In 1970, an eminent political economist wrote that,

> An early, completely forgotten empirical work with a related theme has the significant title *The Triumph of Mediocrity in Business*, by Horace Secrist, published in 1933 by the Bureau of Business Research, Northwestern University. The book contains an elaborate statistical demonstration that, over a period of time, initially high-performing firms will on the average show deterioration while the initial low performers will exhibit improvement.

The author was blissfully unaware of the reason that Secrist's conclusions had no impact. He read Secrist, but overlooked Hotelling.

Similarly, Mark Hirschey, a business school professor and prolific author, wrote that, "Experienced investors know that competitor entry in highly profitable, high-growth industries causes above-normal profits to regress toward the mean. Conversely, bankruptcy and exit allow the below-normal profits of depressed industries to rise toward the mean." It is tempting to believe, as did Secrist and Hirschey, that a company that has a great year must be a great company. If next year is not so great, something must have happened—such as the entrance of new competitors—that undermined its greatness. Experienced investors must find it hard to believe that a great year might be a lucky year.

Oddly enough, Hirschey recognized a firm's success can be affected by luck, but did not consider the possibility that the subsequent reversal might be due to the transitory nature of luck:

> Economic theory explains the often-observed mean reversion in business profits over time as a typical characteristic of the competitive environment. Unexpected good fortune in the form of rising prices or falling costs translates into above-normal profits that act as a magnet for new competition.

This is not an isolated view. Many well-informed observers who have documented regression to the mean in company profits ignore the role of luck and instead believe that there must be an economic explanation—most often, that competitors who see companies thriving will emulate them and erode their profits, while companies doing poorly will change their practices so that they can do better. In a finance book that was required reading for the CFA exam, Robert Haugen wrote that,

> firms quickly *revert to the mean* in terms of the growth rates they report in earnings per share. . . . because [struggling] companies tend to reorganize and reinvent themselves or are taken over and forced to do just that, and because growth companies face hungry competitors eager to participate in profitable product markets.

A investments textbook written by William Sharpe, an economics Nobel laureate, argued that, "ultimately, economic forces will force the convergence of the profitability and growth rates of different firms." To support this assertion, he looked at the firms with the highest and lowest profit rates in 1966. Fourteen years later, in 1980, the profit rates of both groups were closer to the mean. He concluded triumphantly: "convergence toward an overall mean is apparent. . . . the phenomenon is undoubtedly real." *Déjà vu, déjà vu.* Like Secrist 50 years earlier, he did not consider the possibility that this convergence was simply statistical regression.

Several years later, two other distinguished finance professors— one, another economics Nobel laureate—made the very same error. In a lead article in the *Journal of Business*, Fama and French found regression in earnings data and, like Secrist, attributed it entirely to competitive forces:

> In a competitive environment, profitability is mean reverting within as well as across industries. Other firms eventually mimic innovative products and technologies that produce above normal profitability for a firm. And the prospect of

failure or takeover gives firms with low profitability incentives to allocate assets to more productive uses.

There may be some truth in these causal explanations, but they are surely not the whole story. Their evidence is no more persuasive than was Secrist's evidence since they completely ignore the purely statistical explanation that companies with relatively high earnings are likely to have experienced good luck.

Well-run companies are generally more profitable than poorly run companies. True enough, but there is also an element of luck—discoveries of new resources and technologies, the introduction of products that are more successful than expected, fortuitous troubles by competitors, legal battles won and lost. Those companies that have the highest return on assets in any given year are likely to have had good luck—which means that their return on assets was probably higher than the year before and the year after.

Growth Convergence

One of the most fundamental questions in economics is why some countries flourish while others languish. Adam Smith's *The Wealth of Nations*, written in 1776, defied the conventional belief that a nation's wealth should be gauged by its hoard of gold and other precious metals. Instead, Smith argued that a nation's wealth depends on the productivity of its citizens and their ability to utilize their talents to produce goods and services and trade them freely with others. A nation is enriched by people doing what they do best and trading what they make for what they want.

Smith was particularly critical of the mercantilist idea that a nation should use tariffs to discourage its citizens from buying products from other countries and thereby depleting its stock of gold. The argument that candle makers and clothes makers benefit from trading with each other is true across countries as well as within countries. It makes no sense to force a family to wear its own poorly made clothes or to force a country to forgo coffee and bananas because its climate is ill-suited for growing them.

The Wealth of Nations was written in the early years of the industrial revolution and provided an intellectual foundation for the unprecedented economic growth that ensued. Yet, even today, some nations—even nations abundantly rich in natural resources—are desperately poor. Per capita incomes are in the tens of thousands of dollars in industrialized countries (Australia $40,000, France $30,000) and a few thousand or less in undeveloped countries ($2,000 in Nigeria, $500 in El Salvador). These disparities did not arise overnight. They are the cumulative effects of differing growth rates compounded over many decades.

In 1870, real per capita income was $1,800 in England and $1,400 in the United States (30 percent higher in England). One hundred years later, in 1970, real per capita income was $7,000 in England and $11,000 in the United States (60 percent higher in the United States). Over this hundred-year period, the annual growth rate was 2.1 percent in the United States and 1.4 percent in England. The miracle of compound interest multiplied this seemingly small difference in annual growth rates into huge differences in per capita income.

So the question of rich and poor countries is ultimately a question of economic growth. Why do some countries grow faster than other countries for decades or centuries? No one really knows, but there is one consistent characteristic of economic growth across countries. Yep, regression.

The Penn World Table is a data base of annual output in 167 countries, in some cases going back to 1950. There are data for 1970 through 2010 for 143 countries. I divided this forty-year period in two, the first half from 1970 to 1990, the second from 1990 to 2010. Then I calculated the annual growth rate in real per capita output for each country for the first half (between 1970 and 1990), and divided these growth rates into quartiles. The countries in the top quartile had an average annual growth rate of 4.8 percent, the bottom quartile averaged negative 1.6 percent. I then calculated the average growth rate for the countries in these 1970 to 1990 quartiles over the next 20 years, 1990 through 2010. Figure 4 shows the growth rates from 1970 to 1990 and from 1990 to 2010.

Figure 4
Growth Convergence in Real Per Capita Output

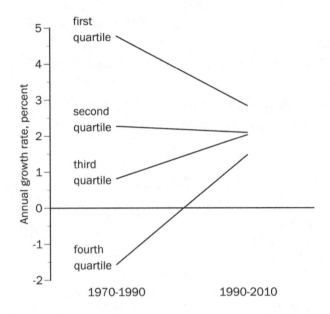

The most striking feature of these data is the convergence in growth rates. Those countries growing the fastest between 1970 and 1990 grew at a slower rate between 1990 and 2010, while those countries growing the slowest between 1970 and 1990 grew faster between 1990 and 2010.

Some economists have labeled this convergence the "middle-income trap," in that fast growth evidently tends to be followed by slower growth and slow growth tends to be followed by faster growth. Nations are apparently sucked into a depressing mediocrity. Perhaps prosperous countries become fat and lazy, while floundering economies get a good kick-start (in some cases, administered by peaceful or violent changes in government). You can probably think of other plausible theories.

There is a simpler explanation. Figure 4 has a remarkable similarity to the business-profits graphs earlier in this chapter because the convergence may be nothing more than statistical regression.

We could create an illusion of divergence from the mean by grouping the countries into quartiles based on their 1990 to 2010 growth rates. Figure 5 shows that those countries growing the fastest between 1990 and 2010 grew at a slower rate between 1970 and 1990, while those countries growing the slowest between 1990 and 2010 grew faster between 1970 and 1990. Just what we would expect from regression.

The similarity between the country growth rates in Figure 5 and the profit rates in Figure 1 is striking, yet a substantial literature on growth convergence developed, written by prominent economists who somehow overlooked regression.

Figure 5
Growth Divergence in Real Per Capita Output

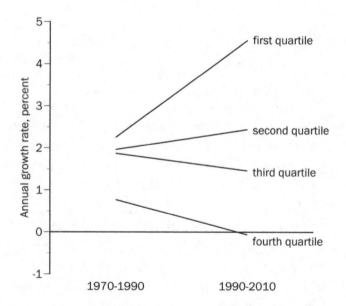

For example, a book and book review, both written by widely respected economists, argued that the economic growth rates of nations converge over time, just like we saw here. They completely ignored the role of regression in this convergence. Milton Friedman wrote an apt commentary titled, "Do Old Fallacies Ever Die?"

I find it surprising that the reviewer and the authors, all of whom are distinguished economists, thoroughly conver-

sant with modern statistical methods, should have failed
to recognize that they were guilty of the regression fallacy.
. . . However, surprise may not be justified in light of the
ubiquity of the fallacy both in popular discussion and in
academic studies.

This regression fallacy doesn't die, because it is easier to think
that companies and countries are governed by predictable economic
forces than to admit how much is due to chance.

Some nations do grow faster than others for decades, but there
is also considerable variation in growth rates for all countries, no matter how fast or slowly they grow, and this variation creates the statistical illusion of a middle-income trap. Countries are not converging
to a depressing mediocrity where every country grows at the same
rate. That is not a question worth studying. The interesting question
remains why some countries grow faster than others for decades.

In 2014 Lant Pritchett and Lawrence Summers, professors at the
Harvard Kennedy School of Government, studied the question and
concluded that we still don't know the answer. Economists have figu-
ratively looked at everything but the kitchen sink, yet, "Nearly every
assertion about correlates (or causes) of growth emerging in any study
has been challenged as not robust in a later study." You shouldn't be
surprised by what does stand out, time and time again, if we only open
our eyes and see it: "Regression to the mean is perhaps the single most
robust and empirical relevant fact about cross-national growth rates."

15

From Bad to Better and Great to Good

N 2010 I GOT A PHONE CALL FROM AN INTERNET COMPANY (I'LL CALL them WhatWorks) that tests web page layouts for other companies. For example, they might compare four different layouts for a client's home page: the current layout and three alternatives that are visually appealing. WhatWorks bases its recommendation on the following experiment. When a potential customer goes to the client's web site, a random-event generator takes the customer to one of four versions of the home page. WhatWorks records the revenue generated by users clicking on ads and, after several days of tests, tells the client which version of the page was the most successful.

This is a well-designed experiment in that it taps the power of random sampling. However, WhatWorks had a recurring problem that they couldn't explain to their clients. After they had made their recommendation and the client followed their advice, the actual benefits generally turned out to be smaller than observed in the experiment. If WhatWorks estimated that the client would see a five percent increase in revenue by switching from its current layout to another layout, revenue might go up only two percent.

Sure enough, it's regression to the mean. For each layout, the benefits fluctuate because of the luck of the draw in who goes to the page, how the user is feeling at that moment, and so on. To the extent there is randomness in this experiment—and there surely is—the actual benefits going forward from the layout that scores highest in the experiment will probably be closer to the mean than were the benefits recorded in the experiment. There is nothing wrong with that. It is still the most successful layout. The client (and WhatWorks) just need to recognize that it is perfectly natural for the benefits to be more modest than predicted by the test.

This is a simple example of how projected benefits can turn out to be smaller than anticipated. The flip side is that apparent benefits can sometimes materialize out of thin air. After one of the WhatWorks data wonks ("Jake") understood regression to the mean, he played a prank on his coworkers.

Jake reported that he had identified 3,000 web sites that were generating very little traffic. He told his coworkers that when he double-clicked on whatever image happened to be in the upper left corner of each page, traffic increased by an average of 800 percent that day. He concluded that,

> The results clearly indicate that double-clicking the image resets each domain and unleashes its true potential. I'll continue watching the results because it appears the effect is already wearing off, so perhaps a periodic refresh is in order. We can simply hire a temp to do it once every few weeks (man, my carpal tunnel is acting up). The domain list is below if you'd like to verify this miraculous result for yourself.

Wouldn't you know it, some of Jake's coworkers tried it, with one reporting that, "All right, I prepped the temp to handle this bi-weekly." Another wondered,

> What is the "image in the top left corner" and why is it clickable? I don't think this a common feature in our designs. This sounds like a bug work-around to me, but I can't reproduce it as described.

Jake fessed up, telling them that it was a joke. He hadn't double-clicked on any images, He hadn't done anything but collect a list of unpopular web sites and count on regression to kick in. If a web site is performing very poorly, it is more likely to be performing below its ability than above its ability, which means that its future performance is likely to increase for no reason other than its performance had been artificially low.

This is a lesson that extends far beyond web-page pranks. Sometimes a company will experience a sharp downturn and bring in highly-paid management consultants to turn things around. The consultants poke around, make some recommendations, and the company miraculously improves.

That's Jake's joke on a larger scale. A company doing poorly is more likely to have been experiencing bad luck than good luck. So, its ability is above its current poor performance and the company is likely to do better in the future, whether or not the consultant's recommendations have any merit or, indeed, whether a consultant is even hired.

Consultants are not inevitably worthless. However, consulting can be similar to temporary ailments that are cured by worthless medical treatments. We need to recognize that inevitable fluctuations in business performance can create an illusion of a problem and a cure when there is neither. In medicine, randomized controlled trials can be used to evaluate treatments, taking fluctuations and regression into account. There is nothing similar in business.

Some CEOs love management consultants, often, I believe, because the consultants give management the cover to do what management wanted to do all along. The consultants are the bad cops, the managers the good cops.

Some observers are skeptical of management consultants, most credibly if they are consultants themselves. In a damning 2006 article in *The Atlantic*, Matthew Stewart wrote about the "The Management Myth," the misplaced faith that there is a coherent theory of management that consultants can use to revitalize companies:

> As a principal and founding partner of a consulting firm that eventually grew to 600 employees, I interviewed, hired, and worked alongside hundreds of business-school graduates, and the impression I formed of the M.B.A. experience was that it involved taking two years out of

your life and going deeply into debt, all for the sake of learning how to keep a straight face while using phrases like "out-of-the-box thinking," "win-win situation," and "core competencies."

I don't know how effective specific management consultants are, but I do know that we can't judge their effectiveness without taking regression into account.

The same is true when there is a change in management. The replacement of a manager when a business is doing poorly is likely to be followed by improved performance regardless of whether the new manager makes a difference. At the other end of the performance spectrum, the hiring of a new manager when a business is doing great is likely to be followed by a dip in performance.

This is clearest in the sports world where wins and losses are a convenient measure of success. In his last twelve seasons at UCLA, John Wooden coached the men's basketball team to ten national championships, including an incredible seven in a row. How could his successor possibly do as well? If Wooden had stayed another five years, he probably wouldn't have done as well.

Wooden wrote in his autobiography that in his final season, after winning his tenth championship, a UCLA fan confronted him as he walked off the court: "Great win coach, this makes up for letting us down last year." Clearly, the fan did not appreciate how difficult it is to win national championships.

Gene Bartow replaced Wooden and his teams won 85.2 percent of their games in two years, compared to Wooden's 80.8 percent, but UCLA didn't win the national championship and Bartow was replaced. Next came Gary Cunningham, whose teams won 86.2 percent of their games in two seasons, but not the national championship, so he was replaced, too. It is a lot easier to replace a loser than to replace a legend. (Think Jack Welch, Walt Disney, Sam Walton.)

It is hard to do better than someone who has been phenomenally successful. It is easy to do better than someone who has done miserably. Because of regression, phenomenal success tends to be followed by diminished success; miserable failure by less failure.

This regression creates a bias for change. Organizations doing great are unlikely to make changes and will typically see their performance regress downward—suggesting that they should have made changes. Organizations doing poorly are more likely to make changes and will typically see their performance regress upward, confirming the value of the changes they made. It will appear that complacency is harmful and change *per se* is beneficial. Once again, it is easy to be fooled by regression to the mean.

Great to Good

There is another unfortunate consequence of ignoring the role of luck in business performance. No matter how we measure success (sales, profits, stock returns), every company has its ups and downs. Performance is consequently an imperfect gauge of ability and susceptible to regression.

The misadventure starts when a researcher compiles a list of top-performing companies and looks for things they have in common, thinking that these common characteristics explain their superior performance. Any group of companies (bad, good, or great) inevitably has some common characteristics. Finding such traits only confirms that we looked and tells us nothing about whether these discovered characteristics are responsible for past successes or are reliable predictors of future success. (What they truly have in common is good luck, but that is not a reliable guarantee of future success.)

Just as medical researchers pillage data, so do business researchers. Ransacking data for patterns is fun and exciting—like playing Sudoku or solving a murder mystery. Examine the data from every angle. Look for something—anything—that is interesting.

This pillaging is known as data mining (aka data grubbing, data dredging, fishing expeditions) and demonstrates little more than a researcher's endurance. We cannot tell whether a data mining marathon demonstrates the validity of a useful theory or the perseverance of a determined researcher. Data without theory is treacherous, and we should be deeply skeptical.

Many management books ransack lists of thriving companies

and then claim to have discovered their secrets to success. In 2001, Jim Collins published his best-selling management book, *Good to Great: Why Some Companies Make the Leap . . . And Others Don't.* It has sold more than four million copies and appeared on several lists of the best management books of all time.

Collins wrote that his book reflects "our search for timeless, universal answers that can be applied by any organization" His conclusion? "We believe that almost any organization can substantially improve its stature and performance, perhaps even become great, if it conscientiously applies the framework of ideas we've uncovered."

Collins and his research team spent five years looking at the 40-year stock market history of 1,435 companies and identified 11 stocks that outperformed the overall market and were still improving 15 years after they made the leap from good to great:

Abbott Laboratories	Kimberly-Clark	Pitney Bowes
Circuit City	Kroger	Walgreens
Fannie Mae	Nucor	Wells Fargo
Gillette	Philip Morris	

Collins scrutinized these 11 great companies and identified five common themes that he labeled with catchy names:

1. Level 5 Leadership: Leaders who are personally humble, but professionally driven to make a company great.
2. First Who, Then What: Hiring the right people is more important than having a good business plan.
3. Confront the Brutal Facts: Good decisions take into account all the facts.
4. Hedgehog Concept: It is better to be a master of one trade than a jack of all trades.
5. Build Your Company's Vision: Adapt operating practices and strategies, but do not abandon the company's core values.

These characteristics are plausible and the names are memorable. The problem is that this is a backward-looking study that is undermined by data mining. Collins wrote that, we developed all of the concepts in this book by making empirical deductions directly from the data. We did not begin this project with a theory to test or prove. We sought to build a theory from the ground up, derived directly from the evidence.

Collins apparently thought this statement made his study sound unbiased and professional. He didn't just make this stuff up. He went wherever the data took him.

In reality, Collins was admitting that he had no idea why some companies do better than others. And he was revealing that he was blissfully unaware of the perils of deriving theories from data. When we look back in time at any group of companies, the best or the worst, we can always find common characteristics. Every one of those 11 companies selected by Collins has either an *i* or an *r* in its name, and several have both an *i* and an *r*. Is the key for going from good to great to make sure that your company's name has an *i* or *r* in it? Of course not.

Finding an *i* and *r* pattern is an obvious example of data mining. Collins' data mining is less obvious because his unearthed patterns sound plausible. It is nonetheless data mining because, as he freely admits, Collins made up his theory after looking at the data.

To buttress the statistical legitimacy of his theory, Collins talked to two professors at the University of Colorado. One said that, "the probabilities that the concepts in your framework appear by random chance are essentially zero." The other professor was more specific. He asked, "What is the probability of finding by chance a group of 11 companies, all of whose members display the primary traits you discovered while the direct comparisons do not possess those traits?" He calculated this probability to be less than 1 in 17 million. Collins concludes, "There is virtually no chance that we simply found 11 random events that just happened to show the good-to-great pattern we were looking for. We can conclude with confidence that the traits we found are strongly associated with transformations from good to great."

It is not clear how this probability of 1 in 17 million was calculated. (I contacted the professor and he couldn't remember.) What is clear is that it is incorrect.

In statistics, this kind of reasoning is sometimes called the Feynman Trap, a reference to the Nobel Laureate Richard Feynman. Feynman asked his Caltech students to calculate the probability that, if he walked outside the classroom, the first car in the parking lot would have a specific license plate, say 8NSR26. Caltech students are very smart, and they quickly calculated a probability by assuming each number and letter were independently determined. This answer is less than 1 in 17 million. When they finished, Feynman revealed that the correct probability was 1 because he had seen this license plate on his way to class. Something extremely unlikely is not unlikely at all if it has already happened.

The calculations made by the Colorado professors and the Caltech students assumed that the five traits and the license plate number were specified before looking at the companies and the cars. They were not, and the calculations are irrelevant.

Collins does not provide any evidence that the five characteristics he describes were responsible for these companies' success. To do that, he would have had to provide a theoretical justification for these characteristics, select companies *beforehand* that did and did not have these characteristics, and monitor their success according to some metric established beforehand. He did none of this.

After the publication of *Good to Great*, not only did these companies regress, some collapsed. Fannie Mae stock went from above $80 a share in 2001 to less than $1 a share in 2008 and delisting in 2010. Circuit City went bankrupt in 2009. Overall, since the book's publication through 2012, five of the 11 great stocks did better than the market, six did worse. A portfolio of those 11 stocks, formed shortly after the book's publication, would have done slightly worse than the market.

In Search of Luck

Twenty years earlier, another best-selling business book did something very similar and had exactly the same problems. Two McKinsey consultants, Tom Peters and Robert Waterman, were asked to study several successful companies. They talked to other McKinsey consultants and came up with a list of 62 leading companies.

In order to make their analysis appear scientific, they looked at six measures of long-term success, three related to growth and three measuring the return on capital and assets. In order to stay in the sample, a company had to rank in the top half of its industry for four of the six measures during the period of 1961 through 1980. As a final screen, they asked industry experts to rate the companies' 20-year record of innovation. The final 43 firms included 35 publicly traded companies and eight companies that were privately held or subsidiaries of other companies.

Peters and Waterman then spoke to managers and read magazine stories about these companies and uncovered eight common traits; for example, a bias for action and being close to the consumer. The book they wrote about their efforts, *In Search of Excellence*, was again a backward-looking study undermined by data mining. There is no way of knowing whether companies with a "bias for action," whatever that means, were more successful than other companies, or whether companies that had been excellent in the past would be excellent in the future.

Michelle Clayman revisited these companies and compared the six measures of long-term success for the five-year period after the book's publication to the five years before publication. In every category, more than two out of three companies did worse after publication. These firms were not jinxed by the publication of *In Search of Excellence*. They regressed because there had been substantial luck in their previously extraordinary performance.

Clayman suggested regression as an explanation for the disappointing performance of these seemingly incredible companies, but her interpretation of regression was incorrect:

> Over time, company results have a tendency to regress to the mean as underlying economic forces attract new entrants to attractive markets and encourage participants to leave low-return businesses.

Regression toward the mean is caused by random fluctuations, not economic forces.

In a follow-up study, Clayman looked at data for 12 years following the book's analysis, and confirmed her initial results. In addition, she looked at the three dozen companies that had the *worst* combination of the six performance measures for the years immediately preceding the book's publication. They generally did better during the twelve years following publication. She concluded, "Almost across the board, the good companies got worse and the poor companies improved." Again, she identified regression, but misinterpreted it:

> There is a phenomenon in nature called reversion to the mean, which asserts that, over time, the properties of the members of the group tend to converge to the average value for the group as a whole. This concept is widely applicable in situations where economic forces tend to move things towards equilibrium.

Not only does this interpretation ignore the role of luck, it sounds suspiciously like Secrist's mistake in believing that companies are converging to an equilibrium of mediocrity. Regression does not assume that "economic forces tend to move things towards equilibrium." Economic forces may exist, but regression is a purely statistical phenomenon that occurs when observed data measure unobserved traits imperfectly; for example, using earnings to measure a company's greatness.

Barry Bannister redid Clayman's initial analysis using a longer time period and a larger list of companies. For every three-year interval during the years 1977 through 1989, Bannister identified the companies that were in the top third for all six measures of success during the previous five years and those in the bottom third. He then compared the stock performance over the three-year interval and concluded that the unexcellent companies generally outperformed the excellent companies. Bannister echoes Clayman's (mis)interpretation:

> [The] key financial ratios of companies tend, over time, to revert to the mean for the market as a whole. The thesis is easily defended. High returns eventually invite new en-

trants, driving down profitability, while poor returns cause the exit of competitors, leaving a more profitable industry for the survivors.

Regression to the mean is a persuasive reason for anticipating that companies that have been extraordinarily successful in the past will be less extraordinary in the future, but this expectation does not depend on competitive forces. This is not to say that competitive forces are a myth, only that we cannot gauge the strength of these forces without taking statistical regression into account.

Longer term, as of 2013 five of the 35 publicly traded companies anointed as excellent by Peters and Waterman have gone bankrupt (Dana Corporation, Delta Airlines, Eastman Kodak, K Mart, Wang Labs). Overall, from the book's publication through 2012, 15 of the Excellent stocks did better than the market, 20 did worse. The average return for a portfolio of the Excellent stocks, formed shortly after the book's publication, would have done slightly worse than the market.

The real lesson from the enduring popularity of such advice is that the authors who write these books and the people who buy them do not realize that the books are fundamentally flawed. This problem plagues the entire genre of books on formulas/secrets/recipes for a successful business, a lasting marriage, living to be 100, so on, and so forth, that are based on backward-looking studies of successful businesses, marriages, and lives.

If we believe that "a bias for action" predicts success, a valid way to test this theory would be to identify companies that have a bias for action and companies that do not, and then see which companies do better over, say, the next ten years. The same is true of secrets for a successful marriage and a long life. Otherwise, we are just staring at the past instead of predicting the future, and we are likely to be fooled by regression to the mean.

Creative Accounting?

A study of the financial statements of hundreds of U.S. companies found that over time, six financial ratios (such as sales to inventory and income to total assets) tended to regress toward the industry averages. The author speculated that industry averages are targets and that when a firm sees a deviation between its ratio and the industry mean, it adjusts its ratios, either by changing its behavior or using the wiggle room provided by generally accepted accounting rules.

Instead of mechanically reporting income and expenses, there is latitude for creative accounting—adjusting the numbers to get the desired results. One joke describes an accountant's job interview:

BOSS:	*How much is 2 + 2?*
ACCOUNTANT:	*4.*
BOSS:	*How much is 2 + 2?*
ACCOUNTANT:	*Uh. . . somewhere between 3 and 5.*
BOSS:	*How much is 2 + 2?*
ACCOUNTANT:	*What do you want me to say?*
BOSS:	*When can you start?*

The vast majority of companies and accountants are honest and well-intentioned—an honesty encouraged by fear of lawsuits, fines, and imprisonment. The ambiguity of accounting practices does give firms wiggle room to aim for industry targets, but we do not have to assign all the blame to creative accounting. Maybe, just maybe, there is a statistical explanation for why financial ratios regress to industry averages.

16

Draft Picks, CEOs, and Soul Mates

I WAS FACULTY CHAIR OF THE ADMISSIONS COMMITTEE AT A VERY SELEC-
tive college for several years, and I know first hand that there is
considerable randomness in which students are admitted and which
are rejected. Mary gets into Harvard and is rejected by Yale; John gets
into Yale and is rejected by Harvard. One of the best students I ever
had applied to the ten best economics PhD programs in the world.
He was accepted by the top program and rejected by the tenth best. I
called a good friend at the tenth-best program and told him that they
had rejected a student who was accepted by the best program. He was
not surprised: "There is a lot of noise in the process."

Because luck plays a role in admissions decisions, we can be confident that the students who are admitted to very selective schools are,
on average, not as good as they appear (and those who are rejected
are typically not as bad as they appear).

This is also true in the real world—life after college. Regression
to the mean is virtually inevitable in hiring decisions based on imperfect information about a person's ability to do the job that the person
is being hired for. Resumes, references, and interviews provide useful,
but incomplete, information. The true ability of the person who appears to be the best of dozens or hundreds of candidates is most likely
not as far above average as he or she seems.

For most jobs, we don't have a good measure of the gap between
what the employers think they are getting and what they actually get.
Sports is one area where we can quantify expectations and reality.

College Football Players

One of the keys to success as a college football coach is recruiting talented players. The culmination of the annual recruiting frenzy is National Signing Day in February, when high school seniors announce the lucky college they have chosen and sign binding letters of intent. The days, weeks, and months leading up to National Signing Day are filled with speculation and rumors. The most sought-after players hold news conferences, televised live by ESPN. With each announcement, some coaches and fans cheer, while others groan.

Rivals.com rates individual players (5-stars is the best, 4-stars the next best) and maintains a widely followed index that rates each college's recruiting class based on the number and quality of signed players. At the end of the National Signing Day, Rivals.com anoints the recruiting champion, runner-up, and also-rans.

The most widely respected metric of actual performance is Jeff Sagarin's rating system based on wins and losses and the relative strength of each team's schedule.

To examine the relationship between the Rivals.com and Sagarin evaluations, we have to take into account the fact that students typically have four years of eligibility, although there may be exceptions for injuries. Players can delay their four years' eligibility by "redshirting," which involves practicing with a team, but not playing in games. (The term redshirt comes from the red jerseys often worn by non-roster players when they practice with the regular players.) A "redshirted freshman" is a sophomore who didn't play his freshman year while a "true freshman" is playing during his first year at college. A fifth-year senior is in his fifth year at college, but the fourth year playing sports, having been redshirted as a freshman.

The idea behind redshirting is that college students get bigger and stronger every year, so that older athletes have an advantage over younger ones. This is particularly true in sports like football, where a year of weight training and physical growth may make a huge difference.

Table 1 shows Sagarin's 2008 performance ratings and Rivals.com's 2004 to 2008 recruiting ratings for the 20 colleges with the highest recruiting averages over this five-year period. The recruiting ratings

use weights to reflect the reality that upperclassmen generally contribute more to a team's success than do underclassmen; for example,
the recruiting class of 2004 is typically more important than the class
of 2005 and certainly more important than the class of 2008 to the
success of the 2008 team.

	Scores		Rankings	
	Rivals.com	Sagarin	Recruiting	Performance
Southern Cal	2835	94.85	1	2
Florida State	2415	83.18	2	10
LSU	2171	81.96	3	11
Oklahoma	2171	94.15	4	3
Miami-FL	2122	77.42	5	14
Florida	2110	98.74	6	1
Georgia	2061	84.81	7	8
Michigan	2055	64.28	8	20
Tennessee	1902	71.95	9	17
Ohio State	1864	84.83	10	7
Texas	1860	93.50	11	4
Alabama	1568	89.48	12	5
Texas A&M	1503	65.14	13	19
Penn State	1439	88.26	14	6
Auburn	1369	71.69	15	18
Nebraska	1356	80.70	16	12
California	1280	83.58	17	9
South Carolina	1106	77.06	18	15
Notre Dame	1106	73.75	19	16
Clemson	967	78.87	20	13

The top-five recruiting teams were, on average, more successful
than the bottom-five teams. However, four of the top-five recruiting
teams underperformed based on their recruiting and all five of
the bottom-five recruiting teams over-performed. It is possible that
coaches who are good recruiters are bad coaches, and vice versa, but
a far more plausible explanation is regression to the mean. There is a
lot of uncertainty in predicting how well high school players will do
playing college football four or five years later. The highest-rated high

school players are more likely to be overrated than underrated. It follows that the best and worst recruiting classes are probably not as far from average as their recruiting scores suggest.

Do the Pros Get it Right?

What about the pros, where hundreds of well-paid scouts have had four years to watch college players compete against each other? In the National Football League (NFL), there is a draft during each offseason in which the teams choose, one by one, players coming out of college. The team with the worst record has the first pick and tries to select the best player available based on scouting reports and an extensive study of the player's college career statistics. Potential draft picks are also invited to an NFL Scouting Combine, a week-long testing ground where players are measured, timed, and quizzed.

First-round picks are not made lightly. An early pick requires a very expensive contract and offers an opportunity to turn a franchise around—maybe not from worst to first, but from worst to respectable. Wasted money on wasted picks cripples a team and can lead to the firing of the people who made the bad picks.

How well does it work? Table 1 shows the first picks for 25 years, 1990 to 2014, and how they worked out. There were some spectacular successes (Peyton Manning and Andrew Luck), but most were disappointing in that their performance was closer to the mean than was their promise. Some were complete flops. Hardcore fans love debating who was the worst first-pick ever—JaMarcus Russell, Steve Emtman, Tim Couch?

If the first player drafted turns out to be the best player his first year in the NFL, he will be the rookie of the year. Only two of these 25 number one picks were rookies of the year. Thirteen went to the Pro Bowl (essentially an all-star team) sometime in their career; twelve never went (though some are young enough that they may go someday). With few exceptions, they turned out to be less successful than players taken later in the draft, sometimes much later.

Table 1

National Football League Number One Draft Picks

Year	Name	Position	College	Pro Team	Rookie of Year	Pro Bowl
1990	Jeff George	QB	Illinois	Indianapolis	no	no
1991	Russell Maryland	DT	Miami (FL)	Dallas	no	yes
1992	Steve Emtman	DT	Washington	Indianapolis	no	no
1993	Drew Bledsoe	QB	Washington State	New England	no	yes
1994	Dan Wilkinson	DT	Ohio State	Cincinnati	no	no
1995	Ki-Jana Carter	RB	Penn State	Cincinnati	no	no
1996	Keyshawn Johnson	WR	USC	New York Jets	no	yes
1997	Orlando Pace	T	Ohio State	St. Louis	no	yes
1998	Peyton Manning	QB	Tennessee	Indianapolis	no	yes
1999	Tim Couch	QB	Kentucky	Cleveland	no	no
2000	Courtney Brown	DE	Penn State	Cleveland	no	no
2001	Michael Vick	QB	Virginia Tech	Atlanta	no	yes
2002	David Carr	QB	Fresno State	Houston	no	no
2003	Carson Palmer	QB	USC	Cincinnati	no	yes
2004	Eli Manning	QB	Ole Miss	San Diego	no	yes
2005	Alex Smith	QB	Utah	San Francisco	no	yes
2006	Mario Williams	DE	N. Carolina State	Houston	no	yes
2007	JaMarcus Russell	QB	LSU	Oakland	no	no
2008	Jake Long	T	Michigan	Miami	no	yes
2009	Matthew Stafford	QB	Georgia	Detroit	no	no
2010	Sam Bradford	QB	Oklahoma	St. Louis	yes	no
2011	Cam Newton	QB	Auburn	Carolina	yes	yes
2012	Andrew Luck	QB	Stanford	Indianapolis	no	yes
2013	Eric Fisher	T	Central Michigan	Kansas City	no	no
2014	Jadeveon Clowney	DE	South Carolina	Houston	no	no

(Mis)Judging Talent

The Scouting Combine was established in 1982. This is a week-long invitation-only evaluation of pro football draft prospects. Table 2 shows some of the scores recorded by the top seven quarterbacks taken in the 2000 draft in the order they were drafted, from Chad Pennington, the first quarterback selected, to Tom Brady, the seventh quarterback selected. Table 1 also shows the 1999 scores of Akili Smith, the first quarterback taken in the 1999 draft.

Table 2
Some NFL Combine Scores for Quarterbacks

	Height (inches)	Weight (pounds)	40-yard dash (seconds)	Vertical Leap (inches)	Broad Jump (inches)
2000					
Chad Pennington	76	229	4.81	33.5	111
Giovanni Carmazzi	75	224	4.74	36.5	119
Chris Redman	75	222	5.37	26.5	98
Tee Martin					
Marc Bulger	74	208	4.97		100
Spergon Wynn	76	229	4.91	34.0	108
Tom Brady	77	211	5.28	24.5	99
1999					
Akili Smith	75	227	4.66	34.0	114

If you want a laugh, Google "Tom Brady" + "scouting combine photo." Brady looks more like an accountant (no offense to accountants) than a football player. His skin is pasty, his muscles modest, his belly soft.

He ran a super slow 5.28 40-yard dash. His 24.5-inch vertical leap was 256th out of the 258 players at the combine. The only players who did worse were a 318-pound offensive tackle who jumped 22.5 inches and 338-pound offensive tackle who jumped 23 inches. Heck, even I could jump 24.5 inches when I was in my 20s, and I was barely an amateur athlete, let alone a professional.

On paper, Brady looked like he hardly belonged at the combine. The scouts thought so, too. He was the 199th player selected in the draft, behind the six other quarterbacks listed in Table 1 for the 2000 draft.

It's not like the New England Patriots knew something the other teams didn't know. The Patriots, like every other team, passed up opportunity after opportunity to draft Brady. When the Patriots got to the sixth round, they debated spending their pick on Brady or another quarterback, Tim Rattay. (Rattay was drafted in the seventh round by the San Francisco 49ers and bounced around the league for several years.)

Even after they drafted Brady, the Patriots didn't know what they had. They made him their fourth-string quarterback, behind Drew Bledsoe and two journeymen. Brady only threw three passes during his rookie season, completing one pass for six yards, but, for a variety of reasons, by the end of the season he had moved up to become the number two quarterback.

At the start of Brady's second season, Bledsoe was injured and Brady became the starter. His first two games were distinctly mediocre. If Bledsoe had been ready to play, Brady probably would have headed back to the bench. But Bledsoe was still hurt, and Brady caught fire, winning ten of the 12 remaining regular season games and leading the Patriots to a Super Bowl championship, where he was named Most Valuable Player (MVP). As of 2015, he has led the Patriots to four Super Bowl championships, and three times been named Super Bowl MVP. In 2010, Brady was selected as the starting quarterback for the NFL 2000s All-Decade Team chosen by the NFL Hall of Fame Selection Committee. Tom Brady's Combine numbers could never have predicted this. As Brady quipped, quarterbacks don't do a lot of vertical leaping.

Table 2 also shows 1999 combine numbers for Akili Smith. By these quantifiable measures, Smith was an incredible athlete, far superior to Brady. The NFL scouts thought so, too. Smith was the third player selected in the 1999 draft. He was a complete bust in the NFL. He is widely regarded as one of the ten worst NFL picks of all time.

Brady is hardly the only overlooked player. Another, equally famous mistake was Joe Montana, a Hall-of-Fame quarterback who also won four Super Bowls and was named Super Bowl MVP three times. One of his intangibles was that he didn't get nervous, no matter how intense the pressure. His nickname was "Joe Cool."

One story about Joe Montana involved Super Bowl XXIII. With 3 minutes and 20 seconds left in the game, the 49ers were on their own 8-yard line and trailing Cincinnati 16-13. The huddle was filled with nervous tension until Montana saw John Candy, a popular comedian, in the stands and blurted out, "Isn't that John Candy?" With that unexpected comment, the players knew that Montana wasn't worried at all. The 49ers drove 92 yards and scored the winning touchdown with 34 seconds left in the game.

How do you measure confidence under pressure and other intangibles? You don't. Before the 1979 NFL draft, Joe Montana tried out, along with other potential NFL quarterbacks, before a group of NFL scouts and was given a rating of 6.5 on a scale of 1 to 9. The highest quarterback rating was 8, given to Washington State's Jack Thompson.

Thompson was the third player selected in the 1979 draft (and the first quarterback selected). His NFL career was a bust and Thompson is now considered one of the 20 worst draft picks in history. Joe Montana was the fourth quarterback selected. He was picked at the end of the third round, the 82nd pick overall. He went on to become one of the greatest quarterbacks in NFL history.

It is hard to describe what Montana, Brady, and other great quarterbacks have, even harder to measure. They are calm under pressure. The can see the field, read defenses, make wise decisions. They call audibles at the line of scrimmage, changing the play when they see the way the defense is lined up. When the play starts and they face 300-pound defenders trying to knock them silly, they decide who to throw the ball to, when to run, when to throw the ball away, all the while scrambling for their lives.

How can you identify these qualities? Montana and Brady were grown men in their early 20s who had played football in great college programs (Montana at Note Dame, Brady at Michigan), and yet the professional football scouts couldn't predict how well they would play against NFL defenses. They couldn't predict that they would become Hall of Fame quarterbacks.

It isn't just quarterbacks. In January 2015, a really smart ESPN radio commentator said that when an NFL team signs a free-agent wide receiver who put up great numbers the year before (like receptions per game and yards per reception), 85 percent of the time the numbers are lower on the new team than they were on his previous team. The commentator said that this was because the team signed the wrong guy; they should have signed the quarterback since wide receivers need quarterbacks who throw balls they can catch. There is truth in that, but it is also true that the regression principle tells us that wide receivers with great numbers will usually see their numbers drop the next year, whether they change teams or not.

Are these examples the exception to the rule or are they the rule? Two business school professors, Cade Massey and Richard Thaler, did a formal study of NFL draft picks for the years 1983 through 2008. They calculated that the chances that a drafted player will turn out to be better than the next player in his position drafted (for example, the first quarterback drafted compared to the second quarterback drafted) is only 52 percent, barely better than a coin flip. Yet, teams pay much more for early draft picks than for later draft picks.

They concluded that the top picks, on average, have better NFL careers than do later picks, but they are paid much more than they are worth in that the differences in the success of draft picks is much smaller than the differences in salaries. Not only that but, leaving salary aside, teams that trade down (for example giving up the first pick in the draft for the fourteenth and fifteenth pick) do much better than teams that trade up. Often, the fourteenth pick turns out to be better than the first pick. Almost always, the fourteenth and fifteenth picks together bring more to a team than does the first pick.

Their explanation is, not surprisingly, regression to the mean. College performance statistics, scouting reports, and Combine numbers are an imperfect measure of how well players will perform in the NFL against teams that are much better than any team they played against in college. Those college players who seem to be far superior to other players usually have NFL careers that are much closer to average.

Massey and Thaler conclude that the first draft pick is a curse, not a blessing. First picks do not perform sufficiently far above average in the NFL to justify their cost. A team with the number one pick would typically do better by trading down for multiple lower picks who, in the aggregate, are more valuable and cost less:

> The irony of our results is that the supposed benefit bestowed on the worst team in the league, the right to pick first in the draft, is only a benefit if the team trades it away. The first pick in the draft is the loser's curse.

Predicting Major League Baseball Success

Danny Goodwin was a power-hitting catcher at Central High School in Peoria, Illinois. In his senior season, in front of dozens of major league scouts who had come to watch him, Goodwin led off the game by hitting a 400-foot home run with a wooden bat. The scouts were impressed. The Chicago White Sox chose Goodwin with the first pick in the 1971 draft, but he turned down their $60,000 offer ($350,000 in 2015 dollars) and took a college scholarship from Southern University in Louisiana, where he was selected by *Sporting News* as College Player of the Year his senior year. The California Angels made him the first pick in the 1975 draft and signed him for $125,000 ($550,000 in 2015 dollars). Goodwin hurt his throwing arm soon after, which hindered his ability to play catcher. Still, he was known more for his hitting than his catching and he played first base, pinch hitter, and designated hitter sporadically for three different teams over seven seasons, with a career total of 736 times at bat (equivalent to a full season and a half) and a lifetime batting average of 0.236. Goodwin was the only player to be the first pick in the MLB draft twice, but his professional career was a bust.

There have been 50 Number 1 MLB picks since the draft was instituted in 1965 through 2014. Only three (Bob Horner, Darryl Strawberry, and Bryce Harper) were Rookie of the Year, even though there are Rookie of the Year awards in both the National League and American League and there were two winners in the National League in 1976 and two winners in the American League in 1979. Four Number 1 picks have retired without ever playing in the major leagues.

The MLB draft currently involves 40 rounds plus compensatory picks for players lost to free agency. By contrast, the NFL draft has only seven rounds. In 2015, 1,215 players were selected in the MLB draft, compared to 256 in the NFL draft. (The National Basketball Association draft is even smaller, with only two rounds and 60 picks.)

The number of MLB draft slots varies from year to year, ranging from 679 in 1975 to 1,738 in 1996. Occasionally a late draft pick will pay off. Albert Pujols, generally considered among the top 50 players of all time, was drafted by the St. Louis Cardinals in the 13th round (pick 402). Mike Piazza, one of the best hitting catchers, was chosen in the 62nd round (pick 1,390) by the Los Angeles Dodgers, mainly as a favor to Piazza's father, who was a long-time friend of Dodger manager Tommy Lasorda. Still, these exceptions are remembered because they are exceptions. Late-round picks are like lottery tickets in that fewer than five percent ever play in the majors.

The baseball draft also differs from the NFL draft in that many of the selections have just graduated from high school and almost every selection is not yet "good to go" and spends several years in a well-organized minor league system involving Triple-A, Double-A, High-A, Low-A, Rookie, and Short-Season leagues. A cynic might speculate that the reason there are so many MLB draft slots compared to the NFL and NBA is that baseball has a fully developed minor league system that needs to be stocked with players for the most promising prospects to play with and against.

Under the collective bargaining agreement that went into effect in 2012, the MLB Commissioner's office assigns a slot value for each pick in the first ten rounds. For the 2015 draft, the first pick was given a slot value of $8,616,900, the second pick $7,420,100, and the third pick $6,223,300. Figure 1 shows the 2015 slot values for the 315 picks in the first ten rounds (10 normal picks for each of the 30 teams plus 15 compensatory picks). After declining sharply for the first five picks, the drop becomes more gradual, ending with a $149,700 slot value for picks 300 to 315.

Figure 1
Slot Values for the First Ten Rounds in 2015

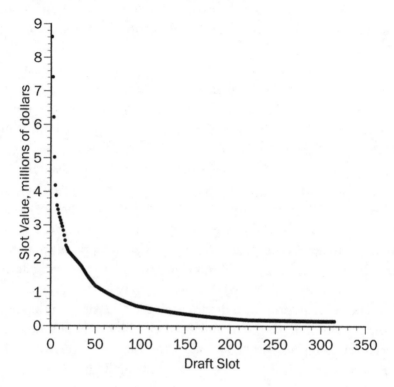

Slot values are benchmarks for signing bonuses, but are not individually binding. Each team has a total bonus pool equal to the sum of their slot values. If a team's total signing bonuses exceeds its allotted bonus pool, it must pay a penalty tax and may also lose future draft picks. For example, a team that goes over its bonus pool by five to ten percent pays a 100-percent tax on the overage and loses its first round pick in the next draft. If a team signs a player for less than his slot value, the difference can be used as part of the bonus pool to sign other players; however, if a team does not sign a player, it loses the slot value. So, there are strong incentives to negotiate signing bonuses close to slot values. In 2015, 17 of the 36 players drafted in the first round signed contracts within 1% of their slot values. However, the top four picks signed for a total of $22,100,000, which was $5,186,600 under their total slot value of $27,286,600.

There is no widely accepted measure of professional success in the NFL, so Massey and Thaler constructed their own metric. In baseball, there is an appealing measure, Wins Above Replacement (WAR), which is an estimate of how many wins a player contributes to his team compared to an easily acquired replacement player—someone who plays the same position in the minor leagues and is not quite good enough to play in the majors.

A WAR of zero means that the player performed no better than an inexpensive replacement. Danny Goodwin's lifetime WAR was –1.7. Even though he had twice been considered the Number 1 prospect, his major league performance was worse than players who were not quite good enough to play in the majors. Still, Goodwin was not that different from many Number 1 picks and luckier than some. Ten percent of the Number 1 picks never played in the majors. Another 12.5 percent played in the majors, but have negative lifetime WARs, and another 12.5 percent have lifetime WARs between 0 and 5.

Since players often spend time in the minors before being moved up to the majors, I analyzed the lifetime WARs through the 2014 season of the top 315 draft picks each year during the 40-year period 1965 through 2004. Table 3 shows the steep decline in MLB performance after the first few draft rounds. Beyond the first few rounds, there is a low probability that a draft pick will play in the majors, let alone excel.

Table 3
Percent of Draft Picks Who Played in Majors and had WARs above 1, 5, and 10

Draft Slot	Played in MLB	WAR > 1	WAR > 5	WAR > 10
1	90.0	72.5	65.0	57.5
2	87.5	67.5	55.0	45.0
3	82.5	55.0	42.5	37.5
4	80.0	52.5	42.5	35.0
5	60.0	37.5	20.0	17.5
10	80.0	45.0	30.0	25.0
50	42.5	17.5	17.5	12.5
150	22.5	7.5	2.5	0.0
315	10.0	5.0	2.5	2.5

Another question is how likely it is that a draft pick will be more successful than players selected later in the draft. It turns out that among the first 315 players drafted each year, 21.0 percent have higher lifetime WARs than the next person drafted, 20.5 percent have lower WARs, and 58.5 percent have equal WARs (mostly because neither played in the major leagues). The same pattern holds when the data are disaggregated into positions; for example, the chances that a right-handed pitcher will do better, worse, or the same as the next right-handed pitcher drafted. The odds are somewhat better for the first 10 draft picks (48.3 better than, 42.8 worse than, and 9.0 equal to the next player drafted) but, after that, a draftee is about equally likely to outperform or underperform the next person drafted. After the first two rounds, the most likely outcome is that neither player makes it to the majors.

Figure 2
Average Lifetime WAR

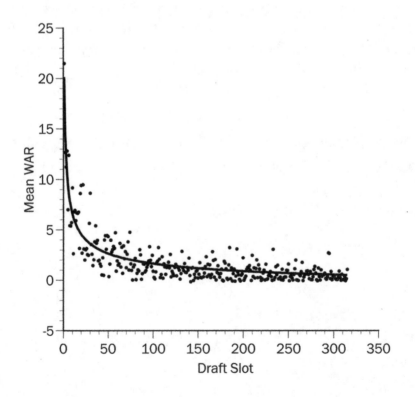

Perhaps, the chances of outperforming the next available player are small, but the payoff is large. For example, 10th picks have outperformed 11th picks 45 percent of the time, underperformed 45 percent of the time, and been equal 10 percent; but maybe the difference in performance has, on average, been substantial when the 10th pick did outperform the 11th pick.

Figure 2 shows the average lifetime WAR for each of the first 315 draft slots. As with earlier comparisons, the differences tend to be largest for first few picks and taper off to very little for later picks. An interesting question is whether this pattern is consistent with the structure of slot values shown in Figure 1.

I calculated the WAR cost by dividing the slot value by the average WAR for every draft slot. The average WAR cost in Figure 3 is $360,000 for one lifetime win above replacement. Most of the first 60 slots (essentially the first two rounds) have an above-average WAR cost, while almost all of the slots after the first three rounds have a below-average WAR cost.

Figure 3
WAR Cost, Slots with WARs < 1 omitted

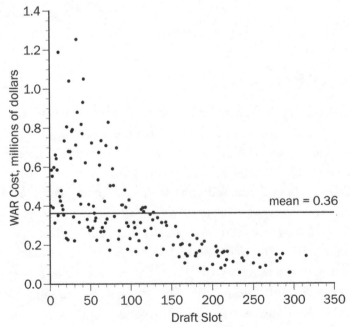

The top picks are overvalued relative to later picks, most likely because of an insufficient appreciation of regression to the mean.

Job Candidates

These principles apply to non-athletes as well as athletes. The reason I focused on football and baseball players is that it is easier to quantify the gap between perception and reality. Predicted athletic success can be gauged by whether a player is drafted first, second, or fourteenth. Actual success can be measured by statistics like a baseball player's WAR or a football player's quarterback rating.

What is comparable for job candidates outside sports? College attended, grade-point-average, job experience, recommendations, and interviews? No matter whether we are talking about a clerk or CEO, there is uncertainty about how well a potential employee will work out.

Is the one person who appears to be the most-qualified out of dozens or hundreds of candidates so extraordinary that he or she had bad luck and still seems the best qualified, or just a somewhat above-average candidate who seems exceptional? The regression principle tells us that the latter is more likely because there are so many merely above-average candidates. Thus, most new hires turn out to be less capable than anticipated when they were hired.

The Peter Principle

The same logic applies to job promotions. Performance in a current job is an imperfect predictor of performance in a higher-level job. It follows that those who are promoted based on their current performance will, on average, be disappointing in their new jobs.

A friend (I'll call him William) works in commercial sales for a large software company. The sales staff is divided into groups, each with a sales manager. Promotions are based solely on each person's sales numbers. Those who make the most sales are promoted to manager, presumably so that they can help those who were not promoted to be more successful salespeople. The problem is that selling something is not the same as training, motivating, and leading people.

William has many vivid examples of people who were great at closing deals, but lousy managers. Some were bosses from hell, bragging about their sales numbers and belittling those who were less successful. Many were well-intentioned, but counterproductive, in that their unhelpful training sessions alienated the sales staff, who felt they were wasting their time. Instead of making sales, they spent far too many hours jumping through hoops and complaining about the hoop-jumping.

When people who get promoted do turn out to be as successful in their new position as they were in their previous position, they are promoted to even higher levels in the hierarchy. And so it goes, until they reach a position where they are not successful and stop getting promoted.

The promotion of people who are good at what they were doing until they are not good at what they are asked to do is the Peter Principle, coined by Laurence J. Peter: "managers rise to the level of their incompetence." People who are promoted based on how well they are doing at their current job, instead of based on the abilities they need to succeed in their new job, will be promoted until they get to a job they are not very good at—their level of incompetence. The Peter Principle is a cynical, but all-too-common, example of regression.

Charismatic CEOs

The same logic applies to CEOs. How can anyone tell whether someone will be a good CEO? We can look at a person's career to date, but the success of a company, unit, or group is never due entirely to a single individual. It's true of the company that the CEO candidate came from and it's true of the company that hires the candidate.

Companies are complex, multifaceted organizations, and the CEO has to deal with employees, suppliers, customers, competitors, and governments. There is enormous uncertainty about how successful a person will be as a CEO. So, there is surely regression in that those who seem to be the best will typically not be as far above average as they seem.

Rakesh Khurana, a Harvard Business School Professor, studied hundreds of CEO firings and hirings and concluded that, in the absence of any reliable quantitative way of predicting success, a company's board of directors is seduced by an unsubstantiated belief that CEOs should be charismatic so that they can inspire employees and placate stockholders.

One of his examples is Eastman Kodak, the legendary camera and film company founded by George Eastman in 1888. (Eastman reportedly added "Kodak" because he wanted a unique, memorable name and he liked the letter *K*.) Kodak grew to become the world's largest photography company. In the 1970s, roughly 90 percent of all cameras and film sold in the United States were made by Kodak. A picture-taking opportunity was known as a "Kodak moment."

However, Kodak was slow to respond to lower-priced competitors and the digital revolution. In 1993, Kodak's board fired its beleaguered CEO, Kay R. Whitmore, a chemical engineer who had worked for Kodak for 36 years, and replaced him with George Fisher, who had been CEO of Motorola, a technology company known for its cell phones and microprocessors. Kodak's stock rose eight percent the day Fisher was announced as Kodak's new CEO.

Figure 4
Kodak's Demise

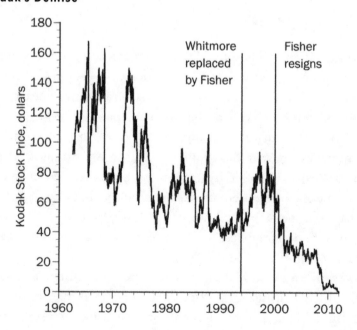

However, Whitmore was not solely responsible for Kodak's weaknesses, nor was Fisher solely responsible for Motorola's strengths. Fisher couldn't save Kodak and he resigned on December 31, 1999, a year before his contract was set to expire. Fisher's charisma kept Kodak stock up for a while, but it didn't last. During Fisher's six years as Kodak CEO, the S&P 500 more than tripled while Kodak rose a meager 6 percent. Figure 4 shows Kodak's stock price from the early 1960s until Kodak filed for bankruptcy protection in January 2012. Whee!

Charisma is rarely the solution to a company's problems. As Warren Buffett put it, "When a manager with a great reputation meets a company with a bad reputation, it is the company whose reputation stays intact."

Disappointment is likely to be especially acute when an outsider is brought in as CEO. First, an outsider doesn't know the company's culture and the strengths and weaknesses of its employees. Second, the board making the hiring decision doesn't know an outsider as well as it knows its own insiders. The less information there is, the more likely the gap between perception and reality—the more likely there will be regression toward the mean.

Searching for Deans

Several years ago, a small private college had a nationwide search for a new college dean. None of the internal candidates were deemed good enough but, from hundreds of external applicants, the search committee identified several who, based on their résumés and references, were invited to interviews held at airports to keep the process confidential. The three candidates who most impressed the search committee were invited to the college for two days of meetings with the faculty, administration, staff, and students.

The search committee was bursting with enthusiasm for these three god-like candidates. However, each candidate came and went, and each was disappointing relative to the pre-visit hoopla. Rumors circulated that the fix was in, that the search committee had deliberately invited two losers in order to make their favorite stand out as the best candidate. However, there was considerable disagreement about which candidate was the committee's favorite.

Do you suspect regression? No one knows how good any of the candidates really are based solely on résumés, references, and interviews. Do you think that the three candidates who look the best are, in reality, even more wonderful than they appear? Or not so spectacular? It would be an extraordinary person who is even better than he or she seems and still manages to rank among the top three candidates. Disappointment is almost inevitable, in that the three candidates who appear to be the best are almost surely not as good as they seem.

Regression also explains why internal candidates are at an inherent disadvantage. Someone who has been at a college for 20 or 30 years does not have very many hidden virtues or warts. Unlike the largely unknown external candidates, with an internal candidate what you see is much closer to what you get.

I discussed this search in my statistics class one year and a student came up to me after class with a remarkable coincidence. The previous night, he had been talking on the phone with his father, who was a sociology professor at another university, and his father was lamenting that when they invite the strongest applicants for faculty positions to come to campus, the candidates are generally not as exciting up close as they seemed on paper and in brief interviews beforehand. The student said he would call his father that afternoon and have a little talk about regression to the mean.

Presidents are Seldom as Good or Bad as They Seem at the Time

When Franklin Roosevelt was running for a third term as U.S. President, George Harvey, the Republican president of the New York City borough of Queens, announced that he would "take the first train for Canada" if Roosevelt were re-elected. Roosevelt won but Harvey stayed in Queens, lamely joking that, "They need me here now more than ever."

In 2010, Rush Limbaugh told radio listeners that if the Obamacare legislation passed "and it's five years from now and all that stuff gets implemented, I am leaving the country. I'll go to Costa Rica." Obamacare did pass and a web site called "A Ticket for Rush" raised money to buy him a ticket to Costa Rica. Like Harvey, Limbaugh stayed in the United States, saying that, "I'll go to Costa Rica for treatment, not move there."

It is not just Republicans who make idle threats to bolt. Shortly before the 2000 presidential election, several celebrities announced that they would leave the United States if George W. Bush became president. Director Robert Altman was very specific, telling reporters that, "If George Bush is elected president, I'm leaving for France." Well, Bush was elected and, like virtually every other celebrity who pledged to flee, Altman stayed. Even though his threat had been filmed, Altman insisted that he had been misquoted: "Here's what I really said. I said that if Bush gets elected, I'll move to Paris, Texas, because the state will be better off if he's out of it." Paris, Texas; Paris, France. What's the difference? It didn't matter much, since Altman didn't move to either.

Maybe Americans aren't very good at geography. During the 2016 election campaign, a digital analytics firm looked at 4.5 million tweets that expressed an opinion about Donald Trump and found that 200,000 included threats to leave the United States. The top destination was Mexico, but 5,800 said they would move to Alaska and another 1,500 said they would move to Hawaii. They knew they didn't like Trump, but they apparently did not know that Alaska and Hawaii are part of the United States.

The deeper issue here is that some Americans get very worked up about presidential elections, a frenzy fed by ambitious politicians trying to convince voters that their opponents are the epitome of evil and will ruin the nation.

Malicious political attack ads are hardly new. In 1800, supporters of Thomas Jefferson distributed pamphlets saying that his opponent, incumbent President John Adams, had a secret plan to declare war on France. Twelve days before the presidential election of 1880, a New York newspaper published a forged letter, purportedly from presidential candidate James Garfield to a fictitious H. O. Morey of the fictitious Employers Union in Lynn, Massachusetts, falsely promising Garfield's support of unlimited Chinese immigration to the United States. Such last-minute shenanigans are so common, they even have a name: the "October surprise" before the November election.

Perhaps the most famous attack ad of all time was during the 1964 contest between Lyndon Johnson and Barry Goldwater, who was allegedly willing to consider using nuclear weapons to end the Vietnam War. In the ad, a two-year-old girl is counting petals as she picks them off a daisy. She reaches nine after a few child-like mistakes, and then a countdown begins in the background that results in a nuclear explosion and a mushroom cloud. The ad ends with a voice-over, "Vote for President Johnson on November 3rd. The stakes are too high for you to stay home." Johnson's campaign put the ad on television only once, but it was shown repeatedly on news and talk shows and was so effective that it is still remembered as the "Daisy ad" more than 50 years later.

A more subtle form of fear-mongering involves push-polling, where people pretend to conduct a political poll, but are really just trying energize citizens to contribute money and vote for their candidate. In the 2000 Republican primary, push-pollsters asked this fake survey question: "Would you be more likely or less likely to vote for John McCain for president if you knew he had fathered an illegitimate black child?" (McCain had adopted a daughter from Mother Teresa's orphanage in Bangladesh.) Another push-poll "survey question" asked voters whether they would be more or less likely to vote for McCain if they knew that the years he spent in a Vietnamese prisoner-of-war camp had made him mentally unstable.

The flip side of those fearing that one candidate will make the country unlivable, are those who hope that another candidate will solve their woes. We are too quick to demonize some and canonize others. Regression to the mean teaches us that, like athletes and CEOs, those presidential candidates who seem the best and worst are seldom as good or bad as they seem at the time. After all, a presidential candidate is just an extreme example of a job candidate.

We can quantify this regression by looking at the public's perception of those who are elected President—presumably with high expectations. Why else did people vote for them?

Since 1937, George Gallup's polling organization has been asking Americans, "Do you approve or disapprove of the way XYZ is handling his job as president?" In the 1970s, several other polling groups have asked the same question. Table 4 shows the results of three surveys conducted in mid-November 2015, regarding President Obama's performance.

Table 4
President Obama's Approval Rating, mid-December 2015

	Approve	Disapprove	No Opinion	Favorability
Gallup	44	52	4	46
ABC/Washington Post	46	50	4	48
CBS	42	47	11	47

The first three columns show the distribution of responses and are roughly comparable; however, the relatively low (42 percent) approval number in the CBS poll is misleading because of the relatively large number of people who declined to answer the question. If we look at the difference between the Approve and Disapprove numbers, Obama actually fares better in the CBS poll (–5) than in the Gallup poll (–8). An attractive way to adjust the numbers is to calculate the approval and disapproval numbers as percentages of the people who answered the question. For instance, the CBS approval number is $42/(42 + 47) = 47$, meaning that 47 percent of the people who expressed an opinion approved of President Obama's performance. These numbers are given in the last column in Table 4, labeled "Favorability," and are very similar. These are the numbers we will look at.

It is interesting to see how presidential approval ratings evolve during a president's term in office. Jimmy Carter defeated the incumbent president, Gerald Ford, in 1976. Carter had graduated from the U.S. Naval Academy and been governor of Georgia. He seemed straightforward and honest, a welcome contrast to some of the presidents who preceded him. During the first month after his inauguration, Carter's favorability was a remarkable 89 percent. By the time he left office in 1981, his favorability was down to 38 percent. Was this an aberration unique to Carter or an example of a general trend for almost all presidents?

Carter is a particularly dramatic example, but it happens to virtually every president. Since 1937, when the polling data begin, nine people have been elected president; all nine had lower favorability ratings at the end of their first term than at the beginning. Five of the six presidents who were elected to two terms had lower ratings at the end of their two terms. (The exception was Bill Clinton who started his first term at 67, finished his first term at 60, and finished his second term at 69.) Three vice-presidents (Truman, Johnson, and Ford) became president after a death or resignation. All three had lower favorability ratings at the end of their terms than at the beginning.

Figure 5 shows the average favorability ratings of the nine elected presidents during their first term in office. The average rating was 82 percent the first month after inauguration, bottomed at 54 percent during their fourth year, and finished at 63 percent—perhaps boosted by their campaigns for re-election or by the public's relief that they will soon be leaving office.

The regression argument suggests that those presidents who have the highest initial favorability ratings will do relatively poorly compared to presidents with more modest ratings. Elected Presidents generally

Figure 5
Average Favorability Rating During First Term in Office

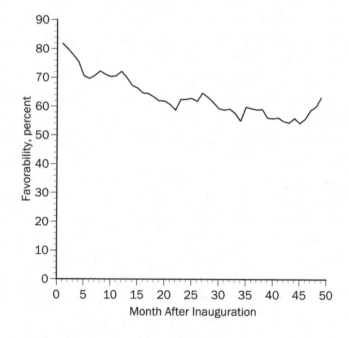

start with favorability ratings above 50 percent; otherwise, how would they have managed to be elected President? So, I divided the nine elected Presidents into those who had above-average or below-average favorability ratings during their first month in office. Figure 6 shows what happened subsequently.

As expected, favorability ratings dropped the most for those presidents who were initially the most popular. In fact, they even dipped below the average favorability rating for the least-popular presidents, before rebounding sharply at the end of their terms. The gap between the average favorability ratings was 17 percentage points (89 versus 72) at the beginning of their terms and 7 percentage points (67 versus 60) at the end.

Popular opinion of presidential candidates is buoyed by speeches, ads, and everything else that goes into presidential campaigns and, also, by our hopes of what they will accomplish if elected. After the election, reality sets in. Presidents are seldom as good as they seem when they are elected. They subsequently regress to the mean.

Figure 6
Presidents with Above-Average and Below-Average Favorability Ratings

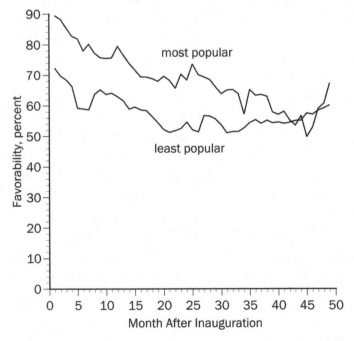

Pizzazz

A woman wrote to the advice column, "Ask Amy":

I keep meeting men who appear to have it all together. Before I know it—it is revealed that they do not. Oftentimes they live with relatives, they are hung up on an ex-wife or girlfriend, they are often financially irresponsible, and/or have serious emotional issues.

Believe it or not, this may just be another example of regression to the mean.

The general principle is that things that seem to be far above or below average probably *are* above or below average, but not as far from average as they seem. One example that resonates with most people is their search for soul mates. Everyone looks for different things, so let's call it "pizzazz." You may be at work or at play when you see someone who seems to have plenty of pizzazz. But when you get to know this person better, there is usually some disappointment. It is possible, but unlikely that those who seem to have the most pizzazz are actually better they seem. How many people could have an off day and still be the most attractive person in the room?

This does not mean that we shouldn't choose those who appear to be the best. What it does mean is that we should be prepared for the likelihood that they are not as great as they initially appear to be. The bad news is that this is just what we should expect. The good news is that we get to keep looking. The sobering news is that the other person probably feels the same way about us.

Once a decision has been made to admit someone, hire someone, or have a serious relationship with someone, be wary of the buyer's remorse that happens when the person turns out to be less than anticipated. Regression explains why the grass is always greener on the other side of the fence and why familiarity breeds contempt. Don't give up on what you have since regression also cautions that what you covet is probably not as good as it seems. Nobody's perfect. Not even you!

VIII. FORECASTING

17

A Better Crystal Ball

STUDY ASKED HUNDREDS OF MEN AND WOMEN TO ESTIMATE THEIR height and weight. The participants were then measured and weighed and the estimates were compared to the measurements. The study found that tall people tended to underestimate their height, while short persons exaggerated their height. People apparently wanted to be more nearly average. The same was true of weight. Heavy persons tended to understate, while thin persons overstated.

The authors of this study concluded that the estimates "converged towards a pair of desired measures." They suggested that, "Our observations also have implications for studies of self-appraisal and body image." Evidently, people didn't want to stand out, sort of like the Japanese proverb, "The stake that sticks out gets hammered down." In Australia, it's called the Tall Poppy Syndrome.

As appealing as that interpretation might be, there is another explanation: regression to the mean. There is an imperfect relationship between our estimates and the actual measurements. Think about it. How much do you weigh at this moment? Your weight fluctuates day to day and even hour to hour and different scales give different measurements. Your estimate of this fluctuating weight is inevitably inaccurate.

To demonstrate the consequences of the imperfect relationship between estimated and measured weight, I made some hypothetical calculations. I assumed that each person's measured weight fluctuates randomly around an average value, at any point in time, being either average or four pounds above or below average. Similarly, I assumed that a person's estimate of his or her weight is either average or five pounds above or below average. For example, if a person's weight averages 160 pounds, measured weight is equally likely to be 156, 160, or 164 pounds and estimated weight is equally likely to be 155, 160, or 165 pounds. The measurement errors and estimation errors are completely independent. People are not biased one way or another.

Finally, I assumed that the average values for the group of people being studied are evenly distributed at five-pound intervals between 140 and 200 pounds. This range is a bit narrow and the uniform distribution is unrealistic, but these simplifications help us focus on the implications. The fluctuations of measurements and estimates may also be a bit high, but I want the consequences to show up clearly in the graphs.

Figure 1

People Want to Believe that They are Close to Average

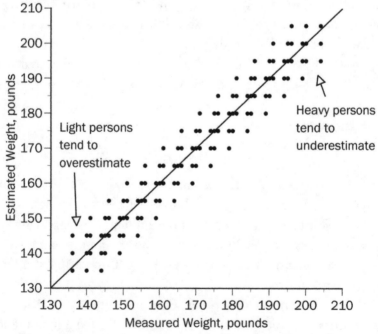

Figure 1 shows the measured and estimated weights for these people. The points above the 45-degree line are people who overestimated their measured weight; the points below the line are those who underestimated. Even though the measurements and estimates are unbiased, there is a pattern to the data. As in the original study, heavy persons tend to underestimate their measured weight, while light persons overestimate their measured weight. The estimated values are closer to the mean than are the measured values, suggesting that people desire to be closer to average than they really are, even though I assumed that every person's estimate is completely unbiased.

Figure 2
People Want to Believe that They are Far From Average

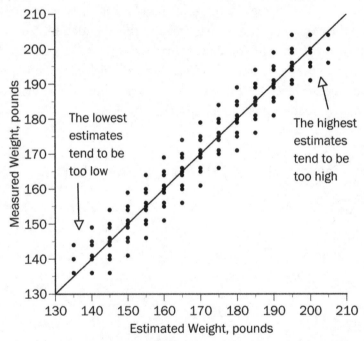

Figure 2 reverses the axes. People above the 45-degree line weigh more than they estimated; people below the line weigh less than estimated. Now, it seems that people overestimate how far they are from average because they don't want to be average! That interpretation is just as wrong as the opposite conclusion, based on Figure 1, that people want to be average. Remember, I assumed that the estimates and measurements are completely unbiased. People may want to be near average, or far from average, but we can't tell from a comparison of estimated and measured values.

This is very similar to our earlier example of the most intelligent women marrying less intelligent men and the most intelligent men marrying less intelligent women. All we are seeing in Figures 1 and 2 is regression to the mean.

Predicting Winners

Before the 2013-2014 National Basketball Association (NBA) season, five widely respected analysts (ESPN, Bleacher Report, Matt Moore, Royce Young, and Zach Harper) predicted the number of wins by each team during the 82-game regular season. Every analyst predicted the Miami Heat to finish first and the Philadelphia 76ers to finish last. For Miami, the predicted number of wins ranged from 57 to 61 games, with an average of 59.8. For Philadelphia, the predictions ranging from 9 to 20 games, with an average of 15. Figure 3 shows the five predictions for each of the 30 teams.

Figure 3
Five Expert Forecasts of 2013-2014 Regular-Season NBA Wins

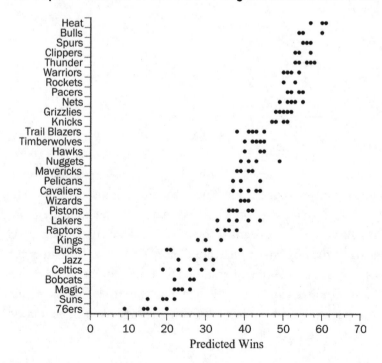

The variation in the expert predictions reflects the uncertainty in game outcomes. The best team doesn't win every game (Miami's predicted 60 wins out of 82 games is 73 percent) so it is not surprising that there is uncertainty about how many games a team will win over the course of a season.

As it turned out, the Heat (predicted to finish first) won 54 games (fifth best in the league) and the 76ers (predicted to finish last) won 19 games (second from the bottom).

Figure 4 shows the imperfect relationship between the actual number of wins and the average prediction of the five analysts. The slope is 0.77, which means that teams predicted to win 10 games more than average tended to win only 7.7 games more than average. The performance of the teams predicted to do the best and worst was closer to the mean than anticipated. The extreme NBA forecasts could have been improved by shrinking them towards the mean.

Figure 4
Average Prediction and Actual 2013-2014 Regular-Season NBA Wins

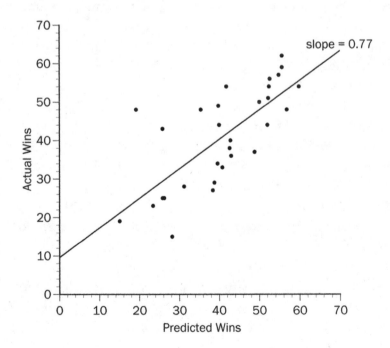

Interest Rate Predictions

Should I take out a fixed-rate or variable-rate mortgage? Will rising interest rates crush the stock market? A lot of people are paid a lot of money to predict changes in interest rates, but interest rates are almost as hard to predict as stock prices. One thing we can be certain of is that the predictions aren't perfect, which suggests that large predicted changes are more likely to be overestimates than underestimates. If so, the accuracy of predicted interest rate changes can be improved by using Kelley's equation to shrink the predictions.

If we interpret Kelley's equation from a Bayesian perspective, a reasonable prior prediction is no change in interest rates. If the predicted changes were always correct, we would use each prediction as is. If predictions were unrelated to actual changes in interest rates, the forecasts would be useless, and we should ignore the forecasts and predict no change. In between these extremes, the Kelley-equation prediction of the change in interest rates is closer to zero than is the expert prediction.

Reid Dorsey-Palmateer and I applied this reasoning to interest rate forecasts from the Survey of Professional Forecasters. The American Statistical Association (ASA) and the National Bureau of Economic Research (NBER) started the ASA/NBER Economic Outlook Survey in 1968. In 1990, the survey was taken over by the Federal Reserve Bank of Philadelphia and renamed the SPF. Approximately 35 professional forecasters are surveyed each quarter. Quarterly forecasts one-to-four quarters into the future are available for the three-month Treasury bill rate, ten-year Treasury bond rate, and Moody's AAA corporate bond yield.

In each case, we used Kelley's equation to adjust the average predicted change in interest rates, based on the historical correlation between predicted and actual changes. Table 1 shows that the adjusted predictions were generally more accurate than the SPF forecasts, though the differences are most persuasive for long-term bonds.

Table 1
Predicting Changes in Interest Rates

Interest rate (months ahead)	SPF	More Accurate Adjusted
T-bill (+1)	37	51
T-bill (+2)	35	50
T-bill (+3)	37	47
T-bill (+4)	36	46
T-bond (+1)	9	37
T-bond (+2)	14	30
T-bond (+3)	15	27
T-bond (+4)	12	28
AAA (+1)	26	62
AAA (+2)	25	61
AAA (+3)	22	62
AAA (+4)	28	54
Total	296	555

The Next Best Thing to Knowing Someone Who is Usually Right

Many financial advisors and portfolio managers use a sophisticated procedure called mean-variance analysis to choose stock portfolios that offer an attractive combination of relatively high expected returns with relatively low risk. Typically, historical data are used to estimate the characteristics of the stocks in the portfolio, and a computer program is used to choose an optimal portfolio.

However, as the warning labels say, "Past performance is no guarantee of future results." Figure 5 shows a scatterplot of the average monthly returns on Dow Jones Industrial stocks for the five-year period 2003 through 2007 and for the subsequent five-year period 2008 through 2012. The correlation is slightly negative, but so close to zero as to be meaningless. The 2003-2007 returns are not reliable predictors of the 2008-2012 returns.

Figure 5
Correlation Between Average Monthly Returns

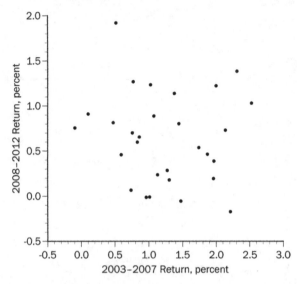

This loose correlation between one five-year period and the next suggests that there is regression to the mean and room for Kelley's equation to guide us to better portfolios. Here, if we account for regression to the mean, we make more accurate forecasts of 72 percent of the average monthly returns for 2008 through 2012.

The same principles apply to asset classes, like U.S. stocks and Treasury bonds. The fact that stocks have done better or worse than bonds over the past 10, 20, or 100 years doesn't mean that the same will be true over the next 10, 20, or 100 years.

One way to handle this reality is to tweak the historical data by changing implausible numbers into plausible numbers. This is, in fact, what Dave Swensen has done in his management of the Yale Portfolio. For example, over the past 90 years or so, U.S. stocks have beaten U.S. Treasury bonds by about eight percent a year, but Swensen expects that, going forward, stocks, on average, will beat bonds by only about four percent a year.

Since Swensen took over in 1985, he has increased the return on Yale's endowment, while reducing the risk. A 2005 cover story for the *Yale Alumni Magazine* was titled "Yale's $8 Billion Man," referring to the fact that Yale's $14 billion endowment would have would have been $8 billion lower if it had earned the same investment returns as the average college and university endowment over the preceding 20 years. In 2013, Yale reported that the value added over the preceding 20 years was $18 billion.

It is not really possible for you or me to emulate Swensen since he invests in gas fields and forests that we can't afford and in hedge funds and private equity deals that we can't access. He also an uncanny knack for picking good managers. Swensen turned down one fund manager who later crashed and burned because the only thing Swensen knew for certain about this man's approach was that he was greedy. Talking about Bernie Madoff, who turned out to be running a massive Ponzi scheme, Swensen said, "If you sat down and had a conversation with him about his investment activities and couldn't figure out that he was being evasive, shame on you."

It has been estimated that some 50 to 80 percent of Yale's success in outperforming other colleges and universities has been due to Swensen's knack for finding managers who outperform the average manager in their asset class, with the remainder due to his use of mean-variance analysis to select a superior asset allocation.

Two students and I applied mean-variance analysis to assets that ordinary people can invest in: a U.S. stock fund, a Treasury bond fund, and a money market fund. We considered three approaches: historical returns, expert opinion, and expert opinion adjusted for regression to the mean.

The historical returns were the past returns on Treasury bonds, U. S. stocks, and money market funds. Our expert opinion came from the Livingston survey. In 1946, a Philadelphia newspaper columnist, Joseph Livingston, started a semiannual survey of business economists' forecasts of macroeconomic variables. The Federal Reserve Bank of Philadelphia started a data base of the survey responses in 1978 and took over the survey after Livingston died. The Philadelphia Fed surveys professional economic forecasters from a wide variety of industries, including non-financial businesses, investment and commercial banks, academic institutions, and government agencies. We looked at the expert forecasts of bond and stock returns.

Our regression-to-the-mean portfolio was based on the argument that when predicted returns are unusually high, these predictions are likely to be too high. For example, if the average stock return has historically been 10 percent and the expert prediction is 20 percent, a better prediction might be closer to the historical average of 10 percent. So, our regression-to-the-mean portfolio shrunk the Livingston predictions towards the historical values.

During the years we studied, annual predicted stock returns ranged from -0.3 percent to 8.8 percent, while predicted bond returns ranged from -1.4 percent to 4.3 percent. These forecasters were surely taking current economic conditions into account, not just using historical averages. It is also clear (and unsurprising) that their forecasts are imperfect. What may be surprising is that their predicted bond returns were essentially uncorrelated with actual bond returns, demonstrating how hard it is to predict interest rates. Their stock forecasts were negatively correlated with actual stock returns. Yep. When these professional forecasters were optimistic, the stock market tended to do poorly; when they were pessimistic, the market tended to do well. Although their forecasts might seem worthless, they are not. It may be as profitable to know forecasters who are usually wrong as to know forecasters who are usually right.

It turned out that the regression-to-the-mean portfolio crushed the portfolio based on the historical data and the portfolio based on the Livingston forecasts. Over the thirteen-year period we studied, an initial investment of $10,000 in a portfolio based on either historical data or the Livingston forecasts grew to $30,000, while the portfolio chosen by the regression model grew to $48,000 (60 percent higher).

Not only can we avoid being misled by regression, we can profit from it. We can out-predict professional forecasters by taking into account the regression they overlook.

IX. INVESTING

18

$100 Bills on the Sidewalk

I N 1996 THE GARDNER BROTHERS WROTE A WILDLY POPULAR BOOK with the beguiling name, *The Motley Fool Investment Guide: How the Fools Beat Wall Street's Wise Men and How You Can Too.* Hey, if fools can beat the market, so can we all.

The Gardners recommended what they called The Foolish Four Strategy. They claimed that during the years 1973 to 1993, the Foolish Four strategy had an annual average return of 25 percent and concluded that this strategy "should grant its fans the same 25 percent annualized returns going forward that it has served up in the past."

Here's their recipe for investment riches:

1. At the beginning of the year, calculate the dividend yield for each of the 30 stocks in the Dow Jones Industrial Average. For example, on December 31, 2013, Coca-Cola stock had a price of $41.31 per share and paid an annual dividend of $1.12 per share. Coke's dividend yield was $1.12/$41.31 = 0.0271, or 2.71 percent.

2. Of the 30 Dow stocks, identify the ten with the highest dividend yields.

3. Of these ten stocks, choose the five with the lowest price per share.

4. Of these five stocks, cross out the one with the lowest price.

5. Invest 40 percent of your wealth in the stock with the next lowest price.

6. Invest 20 percent of your wealth in each of the other three stocks.

As Dave Barry would say, I'm not making this up.

Any guesses why this strategy is so complicated, verging on baffling? Data mining perhaps?

There is a glimmer of logic in Steps 1 and 2 in that the stock market sometimes overreacts to good or bad news, causing stock prices to temporarily be too high or low. These temporary excesses create profitable opportunities for a contrarian strategy of buying what others are selling and selling what others are buying. The first two steps in the Foolish Four strategy (select the ten Dow stocks with the highest dividend/price ratios) is one way of identifying out-of-favor stocks that have low prices relative to their dividends. There is even a long-established investment strategy, called the Dogs of the Dow, that recommends buying the Dow stocks with the highest dividend yields and this sensible strategy has been reasonably successful.

Beyond this kernel of a borrowed idea, the Foolish Four strategy is pure data mining. Step 3 has no logical basis since the price of one share of stock depends on how many shares a company has outstanding. If a company were to double the number of shares, each share would be worth half as much. There is no reason why a Dow stock with more shares outstanding should be a better investment than a Dow stock with fewer shares outstanding. Berkshire Hathaway (which is not in the Dow) has very few shares and consequently sells for a mind-boggling price of nearly $200,000 per share. Yet it has been a great investment.

What about Step 4? Why, after selecting the five stocks with the lowest prices, as if a low price is good, would we cross out the stock with the lowest price? Why indeed.

And Steps 5 and 6? Why invest twice as much money in the next lowest priced stock as in the other three stocks? We all know the answer. Because it worked historically. Period.

Two skeptical finance professors, Grant McQueen and Steven Thorley, knew that one way to unmask a data-mined strategy is to test it with fresh data. If someone makes up a wacky theory to fit a particular set of data, then see how well it does with data that have not been contaminated by data mining. McQueen and Thorley tested the theory with 1949 to 1972 data and found that it didn't work.

They also had another ingenious way of testing the theory. The Gardners formed their Foolish Four portfolios on the first trading day of January each year. If the theory had any merit, it ought to work just as well if the portfolios were formed on the first trading day of July each year. It didn't.

In 1997, only one year after the introduction of the Foolish Four, the Gardners tweaked their system and renamed it the UV4. Their explanation confirms their data mining: "Why the switch? History shows that the UV4 has actually done better than the old Foolish Four." It is hardly surprising that a data-mined strategy doesn't do as well outside the years used to concoct the theory. The Gardners admitted as much when they stopped recommending both the Foolish Four and UV4 strategies in 2000.

The Foolish Four strategy was indeed foolish.

Torturing Data

The underlying reason why it is easy to find data-mined stock market strategies that temporarily work is that there is a substantial random component to fluctuations in stock prices. It is like trying to "predict" coin flips. I flipped a coin ten times and got these results:

H   H   H   T   T   T   T   H   H   T

My data-mined theory is that coins flips go in threes, starting with three heads, then three tails, then three heads, then three tails. My theory is right nine out of ten times! Are you convinced that I can predict coin flips? I hope not.

Let's test my theory with fresh data, ten more coin flips:

H   H   H   H   T   H   H   T   T   H

This time, I got five right and five wrong—decidedly worthless. If you see how I do it, it's completely unimpressive.

With the stock market, the data mining is not out-in-the-open for everyone to see and there is a natural human hopefulness (and greediness) that fuels our willingness to believe that someone can find a way to beat the market. When the Gardners announced that they had found a way to make 25 percent a year, we wanted to believe that it was true.

There are two reasons for skepticism. First, anyone who truly found a way to clobber the market would choose to get rich quickly rather than trying to get rich slowly by selling books for a few dollars each. The second reason for skepticism is that stock prices are determined by investors making voluntary transactions—some buying and others selling—and a stock won't trade at $20 if it is clear that the price will soon be $30. No one would sell something for $20 that could soon be sold for $30. Even if only a shrewd inner circle knew that the price would soon be $30, they would buy shares and more shares, driving the price today up to $30.

When a stock trades for $20, there are an equal number of buyers and sellers, neither side knowing for sure whether the price will be higher or lower the next day or the day after that. The optimists buy and the pessimists sell. Sometimes, the optimists turn out to be right; sometimes, it is the pessimists.

Things that have happened in the past or are expected to happen in the future will already be embedded in stock prices in that buyers and sellers have taken this information into account. For example, during the 1988 presidential election campaign, it was widely believed that the stock market would benefit more from a George Bush presidency than from a Michael Dukakis presidency. Yet on Friday, January 20, 1989, the day of George Bush's inauguration as president, stock prices fell slightly. Bush's impending presidency was old news. Any boost that Bush gave to the stock market happened during the election campaign as his electoral victory became more certain. The inauguration was a well-anticipated non-event as far as the stock market was concerned.

Stock prices will change if the unexpected happens. However, it is impossible to predict the unexpected. Therefore, it is argued, it is impossible to predict changes in stock prices. This argument even has a name—the Random Walk Hypothesis, which holds that stock price changes are unrelated to previous changes, much as coin flips are unrelated to previous tosses, and a drunkard's steps are unrelated to previous steps.

If the Random Walk Hypothesis is true, then changes in stock prices are as hard to predict as coin flips. However, we know that it is easy to find temporary coincidental patterns in past coin flips that are utterly useless for predicting future flips. The same is true of stock prices, which can be tortured for meaningless coincidental patterns.

This argument is a compelling reason for being skeptical of investing strategies that are fanciful (buy stocks when an NFC team wins the Super Bowl) or based on old news (Ford F-150 sales were up last year). It is also a persuasive reason for being dubious of investment strategies like the Foolish Four that are clearly based on tortured data.

There is a story about two finance professors who see a $100 bill on the sidewalk. As one professor reaches for it, the other one says, "Don't bother; if it was real, someone would have picked it up by now." Finance professors are fond of saying that financial markets don't leave $100 bills lying on the sidewalk, meaning that if there was an easy way to make money, someone would have figured it out by now.

This is not completely true. Stock prices are sometimes wacky. During speculative booms and financial crises, the stock market leaves suitcases full of $100 bills on the sidewalk. Still, when you think you have found an easy way to make money, you should ask yourself if other investors have overlooked a $100 bill on the sidewalk or if you have overlooked a logical explanation. Fortunately, there are ways to take advantage of the fact that most investors underestimate the role of luck on the stock market and consequently overlook regression to the mean.

The Nifty Fifty

Decades ago, many investors gauged a stock's attractiveness by its dividend yield—the annual dividend divided by the stock price. A stock that sells for $100 a share and pays an annual dividend of $5 has a dividend yield of five percent. Investors generally ignored the fact that dividends and stock prices usually increase over time, giving investors capital gains in addition to dividends. As late as 1950, the average dividend yield was nearly nine percent, while Treasury bonds paid only two percent interest. Add in capital gains, and stocks were a bargain.

As the value of growth was increasingly recognized in the 1950s and 1960s, rising stock prices pushed dividend yields below bond rates. By the early 1970s, many investors seemed interested *only* in growth, especially the premier growth stocks labeled the Nifty Fifty. This myopia led them to pay what, in retrospect, were ridiculous prices for growth stocks.

The Nifty Fifty was a small group of "one-decision" stocks, companies so appealing that their stock should always be bought and never sold, regardless of price. Among these irresistible stocks were Avon, Disney, McDonald's, Polaroid, and Xerox. Each was a leader in its field with a strong balance sheet, high profits, and double-digit growth rates. No one stopped to consider that these successful companies might, just maybe, have benefitted from good luck.

Investors thought they had found investments that couldn't fail and pushed stock prices upward. For many, the rising prices confirmed the wisdom of their decision. But is a Nifty Fifty stock worth any price, no matter how high? A basic investment maxim is that, "A great company is not necessarily a great stock." No matter how good or bad a company's management, no matter how large or small a company's profits, no matter how bright or bleak a company's prospects, the attractiveness of a company's stock depends on its price. At some price, a great company's stock is expensive; at some price, a lousy company's stock is cheap.

In late 1972, Xerox traded for 49 times earnings, Avon for 65 times earnings, Polaroid for 91 times earnings. Then the party ended. From their 1972 and 1973 highs to their 1974 lows, Xerox fell 71 percent, Avon 86 percent, and Polaroid 91 percent.

There never was an official list of the Nifty Fifty. A 1977 *Forbes* article refers to a Morgan Guaranty Trust list, another 1977 *Forbes* article refers to a Kidder Peabody list. If any group of stocks is clearly the stocks of the Nifty Fifty legend, it is the 24 stocks that appear on both lists. Table 1 shows the Terrific 24 stocks that were on both lists. It also shows how high the price/earnings (P/E) ratios were on December 31, 1972, compared to the 19.1 P/E for the S&P 500.

Table 1

The Terrific 24 Turned Out to be Lousy Investments

	Price/Earnings Ratio December 31, 1972	Annual Stock Return 1973-2001
Polaroid	90.7	-14.7
McDonald's	85.7	10.5
MGIC Investment	83.3	-6.8
Walt Disney	81.6	9.0
Baxter Travenol	78.5	10.1
International Flavors & Fragrances	75.8	5.7
Avon Products	65.4	6.0
Emery Air Freight	62.1	-1.4
Johnson & Johnson	61.9	13.4
Digital Equipment	60.0	0.9
Kresge (now Kmart)	54.3	-1.1
Simplicity Pattern	53.1	-1.5
AMP	51.8	11.2
Black & Decker	50.5	2.5
Schering	50.4	13.2
American Hospital Supply	50.0	12.4
Schlumberger	49.5	10.4
Burroughs	48.8	-1.6
Xerox	48.8	0.9
Eastman Kodak	48.2	1.7
Coca-Cola	47.6	13.2
Texas Instruments	46.3	11.3
Eli Lilly	46.0	13.1
Merck	45.9	14.3

The Terrific 24 turned out to be the Terrible 24. Over the next 29 years, 18 of these 24 stocks underperformed the S&P 500's 12 percent annual return. Figure 1 shows that the stocks with the highest P/Es were the most likely to underperform the market.

Figure 1
Return versus P/E, Terrific 24

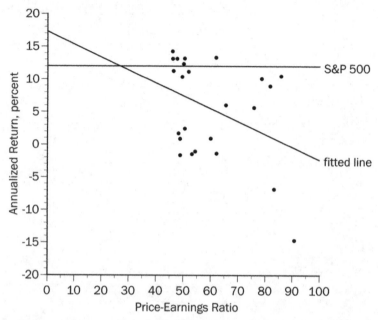

Suppose that someone had invested an equal amount in each of these 24 stocks at the end of 1972, while another investor had invested in the S&P 500. By the end of 2001, the investor who bought the S&P 500 would have twice the wealth of the person who bought the Terrific 24.

The Dangers of Incautious Extrapolation

Growth stocks are often disappointing. Investors appraising a stock with their hearts instead of their minds see temporary growth and think they have found permanent growth. Unfortunately, extraordinary growth is unlikely to be permanent.

The uncritical assumption that an observed trend will continue unabated is incautious extrapolation, which can sometimes be exposed just by thinking through the preposterous implications. A study of British public speakers found that the average sentence length had fallen from 72 words per sentence for Francis Bacon to 24 for Winston Churchill. If this were to continue (famous last words), the number of words per sentence would soon hit zero and then go negative.

Another person, with tongue firmly in cheek, extrapolated the observation that automobile deaths declined after the maximum speed limit in the United States was reduced to 55 miles per hour:

> To which Prof. Thirdclass of the U. of Pillsbury, stated that to reach zero death rate on the highways, which was certainly a legitimate goal, we need only set a speed limit of zero mph. His data showed that death rates increased linearly with highway speed limits, and the line passing through the data points, if extended backwards, passed through zero at zero mph. In fact, if he extrapolated even further to negative auto speeds, he got negative auto deaths, and could only conclude, from his data, that if automobiles went backwards rather than forwards, lives would be created, not lost.

In 1924 the Computing-Tabulating-Recording Company changed its awkward name to something more ambitious—International Business Machines (IBM)—and went on to become the premier growth stock of all time. By 1978, its revenues had been growing (adjusted for inflation) by about 16 percent a year for more than 50 years.

**Figure 2
A Sure Thing?**

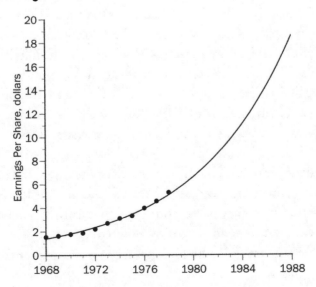

Figure 2 shows a 1978 prediction of IBM's earnings per share over the next ten years, based on an extrapolation of data for the preceding ten years. A smooth line fits these data closely and suggests that IBM's earnings per share would be $18.50 a share in 1988, triple its 1978 earnings. Based on such extrapolations, many stock analysts recommended buying IBM stock, predicting that its price would triple over the next ten years.

Figure 2, like all historical graphs, is merely descriptive. Before we extrapolate a past trend into a confident prediction, we should look behind the numbers and think seriously about whether the reasons underlying the trend will continue or dissipate.

If these analysts had thought about it, they might have realized that IBM's phenomenal growth could not continue forever. IBM started small and grew rapidly as the use of computers spread. Thomas Watson, the CEO of IBM, once said, "I think there is a world market for about five computers." He was wrong. By 1978, IBM was a very large company and there was limited room for continued double-digit growth. It is a lot harder for a giant company to grow by 16 percent a year than it is for a small company.

If IBM kept growing at 16 percent annually and the overall U.S. economy continued growing at its long-run three percent rate, half of U.S. output in 2003 would be IBM products and everything would be made by IBM in 2008! At some point in this fanciful exercise, something has to give—either IBM's growth rate has to drop to three percent or the economy's growth rate has to rise to 16 percent. A sustained 16 percent growth for the entire economy is highly implausible because economic growth is constrained by the labor force and productivity.

Figure 3 shows that IBM's 16 percent growth rate did not persist. A simple extrapolation of the 1968 to 1978 earnings trend proved to be an incautious extrapolation that was recklessly optimistic. Instead of earning $18.50 a share in 1988, IBM earned half that. Subsequent years turned out to be more of the same. IBM could not grow by 16 percent forever and investors who bought IBM stock in the 1970s because they were confident that IBM's remarkable growth rate would never end were disappointed to learn that you seldom see where you are going by looking in the rear-view mirror.

Figure 3
Oops!

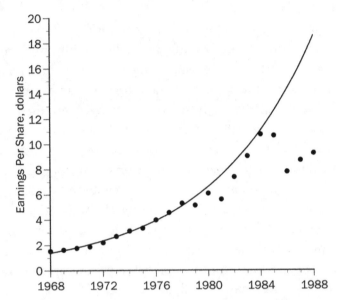

Too often, investors hoping to find the next IBM, Wal-Mart, or Google, see a year or two or three of rapidly increasing earnings and conclude that this is the beginning of years of never-ending growth. Regression teaches us that a company with surging earnings is likely to have experienced good luck and, most likely, will regress towards the mean in the future, disappointing overly optimistic investors.

The same is true of virtually any measure of a company's success. Companies that seem to be the most successful are likely to have benefitted from good luck and will subsequently regress. If investors do not anticipate this regression, stock prices will be too high initially and then wilt when the regression occurs.

Shrunken Earnings Predictions are Better Predictions

Something very similar is true of earnings predictions. The most optimistic predictions are more likely to be overly optimistic than excessively pessimistic. So, the companies with the most optimistic forecasts probably won't do as wonderfully as predicted. Similarly, the most pessimistic predictions are likely to be too pessimistic. So, the companies with the most pessimistic forecasts are likely to beat the gloomy predictions.

Two colleagues and I found that the accuracy of earnings forecasts can be improved consistently and substantially by shrinking them toward the mean. Every year, we looked at the median analyst earnings forecast for the current year and for the year ahead. For example, we looked at the analysts' forecasts that were made in the spring of 1996 for 1996 (the current year) and 1997 (the year ahead). Then we used Kelley's equation to adjust the analysts' forecasts by shrinking them towards the average forecast for all companies.

We didn't analyze the companies' balance sheets. We didn't even look at the companies' names. We just shrank the analysts' predictions towards the mean. We found persuasive evidence that earnings forecasts are systematically too extreme—too optimistic for companies predicted to do well and too pessimistic for those predicted to do poorly. Table 2 shows that, overall, our adjusted forecasts were more accurate seventy percent of the time.

Table 2
Number of More Accurate Predictions

	Analysts	Adjusted
Current-Year Earnings	2,146	5,033
Year-Ahead Earnings	1,264	2,852

If investors pay attention to the analysts (or make similar predictions themselves) stock prices are likely to be too high for companies with optimistic forecasts and too low for those with pessimistic forecasts—mistakes that will be corrected when earnings regress. If this conjecture is correct, stocks with relatively pessimistic earnings predictions may outperform stocks with relatively optimistic predictions.

Five portfolios were formed in the spring of each year based on the analysts' predicted earnings growth rates for the current year. The most optimistic portfolio consisted of the twenty percent of the stocks with the highest predicted earnings growth. The most pessimistic portfolio contained the twenty percent with the lowest predicted growth.

The stock returns were then calculated for each portfolio over the next twelve months. A similar procedure was used for the year-ahead earnings forecasts, but the stock returns were calculated over the next twenty-four months.

Table 3 shows that the pessimistic portfolios trounced the optimistic portfolios. Not only that, the pessimistic portfolios were safer. The most plausible explanation is that the market's insufficient appreciation of regression leaves $100 bills on the sidewalk.

Table 3
Average Percentage Returns for Five Portfolios Based on Analysts' Forecasts

	Most Optimistic	Optimistic	Middle Group	Pessimistic	Most Pessimistic
Current-Year Forecasts	11	15	15	16	18
Year-Ahead Forecasts	23	27	32	35	37

Excessive and Absurd

The great British economist John Maynard Keynes observed that "day-to-day fluctuations in the profits of existing investments, which are obviously of an ephemeral and nonsignificant character, tend to have an altogether excessive, and even absurd, influence on the market." If true, such overreaction might be the basis for Warren Buffett's memorable advice, "Be fearful when others are greedy, and be greedy when others are fearful." If investors often overreact, causing excessive fluctuations in stock prices, it may be profitable to bet that large price movements will be followed by price reversals.

I looked at the daily returns on the stocks in the Dow Jones Industrial Average, from October 1, 1928, when the Dow was expanded from 20 to 30 stocks, through December 31, 2015, a total of 22,965 trading days. I calculated each Dow stock's daily return relative to the average return on the other 29 Dow stocks in order to focus attention on stocks that are buoyed or depressed by idiosyncratic news or opinion specific to an individual company, as opposed to general market surges or crashes caused by macroeconomic news or emotions.

I looked at days when a stock went up or down more than 5 percent and then tracked its return over the next ten days. Stock returns are often assumed to be normally distributed but, in practice, there are many more big days than predicted by the normal distribution. If the normal distribution were correct, there should have been 270 occasions when a stock went up more than 5 percent and another 270 occasions when a stock went down more than 5 percent. In fact, there were 3,810 returns above 5 percent and 3,021 below –5 percent.

What about the aftermath? More often than not, stocks that went up more than 5 percent one day went down the next day, while the reverse was true of stocks going down more than 5 percent. Figure 4 shows that, by the tenth day, there was an average 0.59 percent cumulative loss following positive big days and an average 0.87 percent cumulative gain following negative big days. These are very large annualized returns and highly statistically significant. This is strong evidence of overreaction in that large price changes tend to be followed by persistent, substantial, and statistically persuasive reversals over the next ten days.

Figure 4
Cumulative Average Daily Return After a 5 Percent Big Day

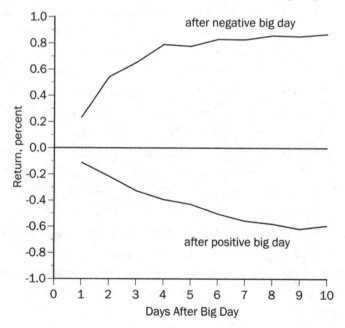

The Curse of the Dow

The Dow Jones Industrial Average (the "Dow") is an average of the prices of thirty blue-chip stocks that represent the United States' most prominent companies. In the words of the Dow Jones company, these are "substantial companies—renowned for the quality and wide acceptance of their products or services—with strong histories of successful growth."

An Averages Committee periodically changes the stocks in the Dow. Sometimes, this is because a stock is no longer traded after a merger with another company. Other times, a company has some tough years and is no longer considered to be a blue chip stock. Such fallen companies are replaced by more successful companies.

For example, Home Depot replaced Sears on November 1, 1999. Sears is a legendary American success story, having evolved from a mail-order catalog that sold everything from watches and toys to automobiles and ready-to-assemble houses, into the nation's largest retailer. Sears had been in the Dow for 75 years, but now was struggling to compete with discount retailers like Walmart, Target, and, yes, Home Depot. Revenue and profits were falling, and Sears' stock price had dropped nearly 50 percent during the previous six months. Home Depot, on the other hand, was booming along with home building and remodeling, and was opening a new store every 56 hours. Its tools competed directly with Sears' legendary Craftsman tools and it was winning the battle. Home Depot's stock price had risen 50 percent in the past six months.

When a faltering company is replaced in the Dow by a flourishing company, which stock do you think does better subsequently—the stock going into the Dow or the stock going out? If you take regression into account, the stock booted out of the Dow probably will do better than the stock that replaces it.

This is counterintuitive because it is tempting to confuse a great company with a great stock. Suppose you find a great company (let's call it LeanMean) with a long history of strong, stable profits. Is LeanMean a good investment? The answer depends on the stock's price. Is it an attractive investment at $10 a share? $100? $1,000? There are some prices at which the stock is too expensive. There are prices at which the stock is cheap. No matter how good the company, we need to know the stock's price before deciding whether it is an attractive investment.

The same is true of troubled companies. Suppose that a company with the unfortunate name Polyester Suits is on a downward death spiral. Polyester currently pays a dividend of $1 a share, but expects its dividend to decline steadily by five percent a year. Who would buy such a loser stock? If the price is right, who wouldn't? Would you pay $5 in order to get a $1 dividend, then 95 cents, then 90 cents, and so on? If the $5 price doesn't persuade you, how about $1? How about 10 cents?

Let's go back to the Dow additions and deletions. The question for investors is not whether the companies going into the Dow are currently more successful than the companies they are replacing, but which stocks are better investments. The stocks going into and out of the Dow are all familiar companies that are closely watched by thousands of investors. In 1999, investors were well aware of the fact that Home Depot was doing great and Sears was doing poorly. Their stock prices surely reflected this knowledge. That's why Home Depot's stock was up 50 percent, while Sears was down 50 percent.

However, regression suggests that the companies taken out of the Dow are generally not in as dire straits as their recent performance suggests and that the companies replacing them are typically not as stellar as they appear. If so, stock prices will often be unreasonably low for the stocks going out and undeservedly high for the stocks going in. When a company that was doing poorly regresses, its stock price will rise; when a company that was doing well regresses, its price will fall. This argument suggests that stocks deleted from the Dow will generally outperform stocks added to the Dow.

Sears was bought by Kmart in 2005, five and a half years after it was kicked out of the Dow. If you bought Sears stock just after it was deleted from the Dow, your total return until its acquisition by Kmart would have been 103 percent. An investment in Home Depot, the stock that replaced Sears, would have lost 22 percent. The Standard & Poor's 500 Index of stock prices during this period was down

14 percent. Sears had an above-average return after it left the Dow, while Home Depot had a below-average return after it entered the Dow. (The Kmart-Sears combination has been ugly, but that's another story.)

Is this comparison of Sears and Home Depot an isolated incident or part of a systematic pattern of Dow deletions outperforming Dow additions? There were actually four substitutions made in 1999: Home Depot, Microsoft, Intel, and SBC replaced Sears, Goodyear Tire, Union Carbide, and Chevron. Home Depot, Microsoft, Intel, and SBC are all great companies, but all four stocks did poorly over the next decade.

Suppose that on the day the four substitutions were made, you had invested $2,500 in each of the four stocks added to the Dow, a total investment of $10,000. This is your Addition Portfolio. You also formed a Deletion Portfolio by investing $2,500 in each of the stocks deleted from the Dow. Table 4 shows how these portfolios did compared to the S&P 500 during the decade following the substitutions. After ten years, the S&P 500 was down 23 percent. The Addition Portfolio did even worse, down 34 percent. The Deletion Portfolio, in contrast, was up 64 percent.

Table 4
The Stocks Added and Deleted on November 1, 1999

	Initial Portfolio	Five Years Later	Ten Years Later
Addition Portfolio	$10,000	$6,633	$6,604
Deletion Portfolio	$10,000	$9,641	$16,367
S&P 500	$10,000	$8,295	$7,652

Still, these are just the four Dow substitutions made in 1999. Maybe 1999 was an unusual year and substitutions made in other years turned out differently? Nope. In 2006, two students, Anita Aurora and Lauren Capp, and I looked at all 50 changes in the Dow back to October 1, 1928, when the Dow 30-stock average began. We found that the deleted stocks did better than the stocks that replaced them in thirty-two cases and did worse in eighteen cases. A portfolio of deleted stocks beat a portfolio of added stocks by about four percent a year, which is a huge difference compounded over 78 years. A $1,000 portfolio of added stocks would have grown to $1.6 million by 2006; a $1,000 portfolio of deleted stocks would have grown to $33 million.

Another way to view the data is to average the returns across stocks, beginning on the day each substitution is made. Thus, we looked at the daily returns for each deleted and added stock for ten years, beginning when the Dow substitution was made. After doing this for all substitutions, we averaged the returns on the first trading day after the substitution, on the second trading day after the substitution, and so on.

Figure 5 shows the ratio of average deletion wealth to the average addition wealth each day over a ten-year horizon. Deleted stocks outpaced added stocks for about five years after the substitution date. Then their relative performance stabilized.

The Dow Deletions have been better investments than the darlings that replace them. Once again, it seems that the market's neglect of regression is leaving $100 bills on the sidewalk.

Figure 5
Ratio of Deletion Wealth to Addition Wealth after Substitution

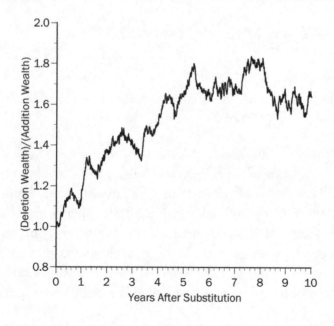

Choosing an Advisor

Because stock picking involves more than a little luck, there is regression in that the investment advisors who make the best stock picks in any given year will, on average, be more mediocre the next year.

In his best-selling, prize-winning book, *Against the Gods*, Peter Bernstein (1996) wrote that,

> The track records of professional investment managers are also subject to regression to the mean. There is a strong probability that the hot manager of today will be the cold manager of tomorrow, or at least the day after tomorrow, and vice versa . . . . [T]he wisest strategy is to dismiss the manager with the best track record and to transfer one's assets to the manager who has been doing the worst; this strategy is no different from selling stocks that have risen the furthest and buying stocks that have fallen furthest.

Bernstein is wise, but this is not wisdom. The idea that the best will be worst and the worst will be best is the gambler's fallacy that good luck must be offset by bad luck. It is false and it is not regression to the mean.

Regression arises because the managers with the best track records probably benefited from good luck and are consequently not as far above average as they seem. Next year, they are likely to be closer to average—not below average. If there is some skill in stock picking, the most successful investment managers can be expected to outperform the least successful ones, but not by as much in the future as they have in the past. If there is no skill, just luck, we may as well pick managers randomly—or save money by not using a manager at all—but there is no reason to choose the worst manager.

X. CONCLUSION

19

Living With Regression

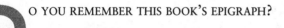D O YOU REMEMBER THIS BOOK'S EPIGRAPH?

> There are few statistical facts more interesting than
> regression to the mean for two reasons. First, people
> encounter it almost every day of their lives. Second,
> almost nobody understands it. The coupling of these
> two reasons makes regression to the mean one of the
> most fundamental sources of error in human judgment.

I've tried to persuade you of the wisdom embraced by this
quotation. We encounter regression almost daily. Yet, we seldom
recognize it, are often surprised when it occurs, and frequently draw
erroneous conclusions. We can do better.

The logic of regression is simple, but powerful. Our lives are
filled with uncertainties. Even death and taxes are uncertain, if we try
to predict when death will occur or how much taxes we owe. The difference between what we expect to happen and what actually does
happen is, by definition, unexpected. We can call these unexpected
surprises chance, luck, or some other convenient shorthand. The
important point is that, no matter how reasonable or rational our
expectations, things sometimes turn out to be higher or lower, larger
or smaller, stronger or weaker than expected.

Even though we experience this over and over again, we are inclined to discount the role of luck in our lives—to believe that suc-

cesses are earned and failures are deserved. We misinterpret the temporary as permanent and invent theories to explain noise. We overreact to the unexpected, and are too quick to think what was previously unexpected is now to be expected.

We see a golfer win the British Open, conclude that he is the best golfer in the world, and expect him to win the next tournament. We see a student get the highest score on a test, conclude that she is the best student in the class, and expect her to get the highest score on the next test. We see a worrisome medical test result, conclude that the patient has a disease, and prescribe a treatment.

In fact, the golfer may have been lucky; the student may have been lucky; the patient may have been unlucky. The thing about luck—good or bad—is that it cannot be counted on to repeat, and the more extreme the luck, the less likely it is to be repeated. Then when it is not repeated, we are tempted to overreact again by inventing a seemingly plausible explanation for the inexplicable.

If the British Open champion loses the next tournament, we might conclude that he wasn't focused. If the student with the highest test score does not do as well on her next test, we might conclude that she did not study as much. If the patient with the worrisome medical result fares better a month later, we might conclude that the prescribed treatment was effective.

If, instead, we recognize that luck may have played a role, we are less likely to overreact. We will realize that the person who wins a tournament is not necessarily the best player, that the student with the highest test score is not necessarily the best student, that the patient with a troubling medical reading does not necessarily have a disease.

We will understand that several golfers are good enough to win a tournament and several students are good enough to get the highest test score, and they take turns doing so—not because their abilities yo-yo week to week, but because their luck comes and goes. We will understand that medical test results fluctuate even if the patient's condition does not.

The key to not being deceived by regression is to look behind the luck—to recognize that when we see something remarkable, luck was most likely involved and, so, the underlying phenomena is probably not as remarkable as it seems.

Whenever there is uncertainty, people may make flawed decisions based on an insufficient appreciation of regression. The examples in this book are from many different aspects of our lives because regression is pervasive. It happens in parenting, education, games of chance, sports, medicine, business, investing, and more.

There is regression in inherited traits such as height, weight, and intelligence because the measurements we see are noisy indicators of the genetic influences that pass from one generation to the next. Regression tells us that abnormal parents generally have less abnormal children, and abnormal children typically have less abnormal parents. A regression fallacy would be to assert that parents who are intellectually or athletically gifted stunt their children, or that parents who are intellectually or athletically challenged stimulate their children.

There is regression in education because academic performances that are far from the mean generally represent talents that are closer to average. Thus, students with remarkably high or low scores can be expected to score closer to the mean on another test of the same material, even if there is no change whatsoever in their mastery of the material. A regression fallacy is to misinterpret changes in tests scores as changes in ability rather than fluctuations in scores about ability. If the students with the highest scores in second grade get somewhat lower scores in fourth grade, it doesn't mean the school failed them. If the students with the lowest scores do better after special tutoring or being screamed at, it doesn't mean that the tutoring or screaming was effective.

There is regression in dice, cards, and other games of chance. Extraordinary performances will most likely be followed by less extraordinary performances, but we should not be surprised and we need not invent excuses. It is a regression fallacy to misinterpret fortune and misfortune as evidence of hot and cold streaks (winning makes winning more likely) or the law of averages (winning makes losing more likely). Hot streaks and the law of averages are both wishful thinking.

A related mistake is to ignore sunk costs—regrettable decisions that cannot be undone. For example, some people act recklessly after a big loss in poker or the stock market, hoping to erase the loss quickly.

For poker players and investors who have sound strategies, the regression principle counsels that patience is better than a Hail Mary.

There is regression when we use athletic performances to assess athletic skills in that performances that are far from the mean exaggerate how far skills are from the mean. A player or team that accomplishes something exceptional most likely benefited from good luck and will subsequently regress. A regression fallacy would be to believe that champions choke or are jinxed by appearing on the cover of *Sports Illustrated*, by a teammate or announcer talking about how well they are doing, or by a fan watching them on television.

There is regression in medicine because of natural variations in diagnostic tests. If a test result is abnormally high or low, a second test will probably yield a result closer to the mean. A failure to recognize this regression can lead to unnecessary treatments and an unfounded belief that worthless treatments are effective.

There is regression in well-designed trials of medications because there is randomness in selection of subjects. On average, one out of every twenty worthless treatments that are tested will show statistically significant effects. In addition, some studies ransack data to find patterns and then make up explanations afterward. If the reported positive results are due to random variations exploited by multiple tests or tortured data, it is no surprise that there is often a decline effect and a medical never-mind as the results regress to the mean.

There is regression in business because of chance fluctuations in most measures of success. It is a regression fallacy to see this regression and conclude that businesses are converging to a depressing mediocrity. Regression in business performance can also mislead us into thinking that management consultants and management changes are the reason for the regression. As with the illusion that temporary physical ailments are cured by worthless medical treatments, temporary fluctuations in business performance can create an illusion of a problem and a cure when there is neither.

Another unfortunate consequence of ignoring the role of luck in business performance is the tendency to look backward at companies that have done well, find a pattern, and conclude that this pattern is the secret to business success. The true common pattern is that these companies were lucky. Going forward, they generally regress.

There is regression in the evaluation of job candidates, no matter whether they are clerks, CEOs, or politicians. Whenever there is uncertainty about how well a person will do on the job, those who seem the most qualified will most likely not do as well as anticipated. It is also true of searches for soul mates.

There is regression in the stock market because investors overreact to corporate news. Unusually strong or weak earnings tend to regress and the most optimistic and pessimistic earnings forecasts are generally too extreme. So, stock prices are often too high for companies whose earnings have increased dramatically or are predicted to do so, while the reverse is true of companies with weak earnings or pessimistic forecasts. When earnings turn out to be closer to average than they have been or were predicted to be, stock prices adjust. It is a regression fallacy to overreact to business news. An appealing alternative is a contrarian strategy—avoiding companies that have done well or are predicted to do well and investing in out-of-favor stocks.

In the same way, stocks that drop sharply are often better investments than are stocks that surge, and stocks that are removed from the Dow are generally better investments than the stocks that replace them. This is counterintuitive for most investors, which leaves $100 bills on the sidewalk for those who understand regression.

Once we recognize the ubiquitous role of luck in our lives, we can anticipate regression before it happens and recognize it for what it is when it occurs. We don't need to invent flimsy theories to explain chance variation. Not only can we stop being fooled by regression, we can take advantage of it.

Regression is a package of pitfalls and opportunities. Avoid the pitfalls and seize the opportunities.

BIBLIOGRAPHY

Bibliography

Abramson, Dan, "A Ticket For Rush: Website Raises Money To Banish Limbaugh," HuffPost Comedy, May 25, 2011.

Anderson, J. and G. Smith, 2006, "A great company can be a great investment," *Financial Analysts Journal*, 62 (4), 86-93.

Anginer, D. and M. Statman, 2010, "Stocks of admired and spurned companies," *Journal of Portfolio Management*, 36 (3), 71-77.

Antman, EM, "Early administration of intravenous magnesium to high-risk patients with acute myocardial infarction in the Magnesium in Coronaries (MAGIC) Trial: a randomised controlled trial," *Lancet* 2002; 360:1189-1196.

Antunovich, P., Laster, D., and S. Mitnick, 2000, "Are high-quality firms also high-quality investments?," *Current Issues in Economics and Finance*, 6 (1), 1-6.

Associated Press, "Drug May Help the Overanxious on S.A.T.'s," *New York Times*, October 22, 1987.

Associated Press, "Seahawks Battle Jinx of the Super Bowl Winner," *New York Times*, January 3, 2015.

Aurora, Anita, Capp, Lauren, and Gary Smith, 2008, "The Real Dogs of the Dow," *The Journal of Wealth Management*, 10 (4), 64-72.

Bannister, B. B. 1990. "In Search of Excellence: A Portfolio Management Perspective," *Financial Analysts Journal*, 46 (2), 68–71.

Barrett, JFR, Jarvis, GJ, Macdonald, HN, Buchan, PC, Tyrrell, SN, Lilford, RJ, 1990, "Inconsistencies in clinical decision making in obstetrics," *Lancet* 336: 549-551.

Baseball-Reference.com, "WAR Explained," retrieved August 8, 2015, from http://www.baseball-reference.com/about/war_explained.shtml

Baum, Gabriel, and Gary Smith, 2015, "Great Companies: Looking for Success Secrets in All the Wrong Places," *Journal of Investing* 24, 61-72.

Baumol, W.J., S.A.B. Blackman, and E.N. Wolff. 1989, *Productivity and American Leadership: The Long View*, Cambridge and London: MIT Press, 1989.

Begley, C. Glenn, & Lee M. Ellis, 2012, "Drug development: Raise standards for preclinical cancer research," *Nature*, 483, 531-533.

Bem, D. J. (2000), "Writing an empirical article," in R. J. Sternberg (Ed.), *Guide to Publishing in Psychology Journals* (pp. 3–16), Cambridge, England: Cambridge University Press.

Bernstein, P. L. 1996, *Against The Gods: The Remarkable Story of Risk*, New York: John Wiley & Sons.

Bland, J. Martin, and Douglas G. Altman, 1994, "Regression towards the mean," *British Medical Journal*, 308: 1499.

Bland, J. Martin, and Douglas G. Altman, 1994, "Some examples of regression towards the mean," *British Medical Journal*, 309: 780.

Brodeur, Paul, *The Great Power-Line Cover-Up: How the Utilities and the Government Are Trying to Hide the Cancer Hazard Posed by Electromagnetic Fields*, Boston: Little, Brown, 1993.

Brodeur, Paul, "Annals of Radiation: Calamity on Meadow Street," *The New Yorker*, July 9, 1990, 66, 38-72.

Brodeur, Paul, "Department of Amplification," *The New Yorker*, November 19, 1990, 134-150

Brodeur, Paul, "Annals of Radiation: The Cancer at Slater School," *The New Yorker*, December 7, 1992, 68, 86-119.

Brown M. Craig, 1982, "Administrative succession and organizational performance: The succession effect," *Administrative Science Quarterly*, 27: 1-16.

Buffett, Warren E., 2008, "Buy American. I Am," *New York Times*, October 16.

Carnevale, A. 1999. Strivers. *Wall Street Journal*, August 31.

Chemi, Eric, "Is Warren Buffett Approaching an All-Time Losing Streak?," *Bloomberg Businessweek*, October 16, 2013.

Clayman, M. 1987, "In Search of Excellence: The Investor's Viewpoint," *Financial Analysts Journal*, 43 (3), 54–63.

Clayman, M., 1994, "Excellence: Revisited," *Financial Analysts Journal*, 50 (3), 61-65.

Cleary, T. A., Humphreys, L. G., Kendrick, S. A., & Wesman, A., "Educational uses of tests with disadvantaged students," *American Psychologist*, 1975, *30*, 15-41.

Clotfelter, C. T., Cook P. J., 1993. "The 'gambler's fallacy' in lottery play," *Management Science* 39 (12) 1521-1523.

"Coffee Nerves," *Time*, March 23, 1981, Vol. 117 Issue 12, p77.

Cohen, Michael D., and James G. March, 1986, *Leadership and Ambiguity: The American College President*, 2nd ed. Boston: Harvard Business School Press.

Cole, P., 1971, "Coffee-drinking and cancer of the lower urinary tract," *Lancet*, 297:1335-1337.

Collins, J., 2001, *Good to Great*, New York: HarperCollins.

Coval, J. D., T. Shumway, 2005, "Do behavioral biases affect prices?" *Journal of Finance* 60 (1) 1-34.

Crandall, Virginia C., 1979, "Young Adulthood Study 1939-1967," Henry A. Murray Research Center, Cambridge, MA.

Crissey, O. L., "Mental development as related to institutional and educational residence," University of Iowa Studies in Child Welfare, 1937, 18, No. 1, p. 81.

Crum, R. L., D. J. Laughhunn, J. W. Payne. 1981, "Risk-seeking behavior and its implications
for financial models," *Financial Management* 10 (5) 20-27.

Cummings S., Palermo B., Browner W., et al. "Monitoring osteoporosis therapy with bone densitometry: misleading changes and regression to the mean," *Journal of the American Medical Association*, 2000; 283 (10):1318–1321.

De Bondt, W. F. M. and R. Thaler, 1985, "Does the stock market overreact?," *Journal of Finance*, 40 (3), 793–805.

De Bondt, W. F. M. & Thaler, R. H., 1987, "Further evidence on investor overreaction and stock market seasonality," *Journal of Finance*, 42 (3), 557–581.

Derry S., Loke YK, "Risk of gastrointestinal hemorrhage with long term use of aspirin: meta-analysis," *BMJ* 2000; 321:1183-87

Dickinson, Amy, 2015, "Ask Amy," *Los Angeles Times*, January 16.

Dorsey-Palmateer, Reid and Gary Smith, 2007, "Shrunken Interest Rate Forecasts are Better Forecasts," *Applied Financial Economics*, 17, 425-430.

Dorsey-Palmateer, Reid and Gary Smith, 2006, "Regression to the Mean in Flight Tests," unpublished.

Durslag, Melvin, 1991, "After 51 Years, It's Time to Say Goodbye," *Los Angeles Times*, May 22.

Educational Records Bureau, 2004, "Comprehensive Testing Program 4: Technical Report," New York: Educational Testing Service.

Eichengreen, Barry; Park, Donghyun; Shin, Kwanho, 2013, "Growth Slowdown Redux: New Evidence on the Middle Income Trap," *NBER* Working Paper No. 18673.

Eichengreen, Barry; Park, Donghyun; Shin, Kwanho, 2012, "When Fast Growing Economies Slow Down: International Evidence and Implications for China," *Asian Economic Papers*, 11, 42-87.

Egger M, Davey Smith G, Schneider M, Minder C. "Bias in meta-analysis detected by a simple, graphical test," *BMJ*, 1997 Sep 13; 315(7109): 629-34.

Elder, R.F., 1934, "Review of The Triumph of Mediocrity in Business by Horace Secrist," *American Economic Review*, 24 (1), 121–122.

Fama, Eugene F., and Kenneth R. French, 2000, "Forecasting Profitability and Earnings," *The Journal of Business*, 73 (2), 161-175.

Farmer, Sam, "Carroll could join another elite club," *Los Angeles Times*, January 20, 2015.

Feenstra, Robert C., Robert Inklaar and Marcel P. Timmer (2013), "The Next Generation of the Penn World Table," available for download at www.ggdc.net/pwt

Feinstein, AR, Horwitz, RI, Spitzer, WO, and Battista, RN, 1981. "Coffee and Pancreatic Cancer: The Problems of Etiologic Science and Epidemiologic Case-Control Research," *Journal of the American Medical Association*, 246: 957-961.

Fesenmaier, Jeff, and Gary Smith, 2002, "The Nifty-Fifty Re-Revisited," *Journal of Investing*, 11 (3), 86–90.

Freedman, David H., "Lies, Damned Lies, and Medical Science," *The Atlantic*, November 2010, 76-86.

Furby, Lita, 1973, "Interpreting Regression toward the Mean in Developmental Research," *Developmental Psychology*, 8 (3), 172-179.

Friedman, M. 1992. "Do Old Fallacies Ever Die?," *Journal of Economic Literature*, 30 (4), 2129–2132.

Gaby, Alan R., "Editorial: Intravenous Magnesium for Acute Myocardial Infarction: The Controversy Continues," Townsend Letter for Doctors & Patients, April 2003.

Galek, Jeff, LeBoeuf, Robyn A., Nelson, Leif D., and Joseph P. Simmons, 2012, "Correcting the Past: Failures to Replicate Psi," *The Journal of Personality and Social Psychology*, 103 (6), 933-948.

Galton, Francis, 1869, *Hereditary Genius*, London Macmillan.

Galton, Francis, 1877, "Typical Laws of Heredity," *Nature*, 15 (388), 492-495; 512-514; 532-533.

Galton, Francis, 1886, "Regression Towards Mediocrity in Hereditary Stature," *Journal of the Anthropological Institute of Great Britain and Ireland*, 15, 246-263.

Galton, Francis, 1889, *Natural Inheritance*, New York: Macmillan.

Gammons, P., 1989, "Inside Baseball," *Sports Illustrated*, 103, 68.

Gardner, D. and T. Gardner, 1996, *The Motley Fool Investment Guide: How the Fools Beat Wall Street's Wise Men and How You Can Too*, New York: Simon and Schuster.

Gardner, D. and T. Gardner, 2000, "Farewell, Foolish Four," retrieved August 28, 2013, from http://www.fool.com/ddow/2000/ddow001211.htm

Garvey, R., A. Murphy, and F. Wu. 2007, "Do losses linger? Evidence from proprietary stock traders," *Journal of Portfolio Management*, 33 (4) 75-83.

Gawande, Atul, "The Cancer-Cluster Myth," *The New Yorker*, February 8, 1999, 34-37.

Greve, Henrich, R., 1999, "The Effect of Core Change on Performance: Inertia and Regression toward the Mean," *Administrative Science Quarterly*, 44 (3), 590-614.

Haugen, Robert A., 1995, *The New Finance*, Upper Saddle River, New Jersey, Pearson Prentice Hall.

Hirschey, M. 2003, *Tech Stock Valuation: Investor Psychology and Economic Analysis*, Amsterdam, Boston: Academic Press.

Hirschman, A.O. 1970. *Exit, Voice, and Loyalty,* Cambridge, Massachusetts: Harvard University Press, 11.

Hotelling, H., 1933, "Review of *The Triumph of Mediocrity in Business,* by Horace Secrist," *Journal of the American Statistical Association*, 28 (184), 463–465.

Hsieh CC, MacMahon B, Yen S, Trichopoulos D, Warren K, Nardi G., "Coffee and Pancreatic Cancer (Chapter 2)," *New England Journal of Medicine* 1986; 314:587–589.

Humphreys, Lloyd G. 1978, "To Understand Regression From Parent to Offspring, Think Statistically," *Psychological Bulletin*, 85, No. 6, 1317-1322.

Hwang, Margaret, Keil, Manfred, and Gary Smith, 2004, "Shrunken Earnings Predictions are Better Predictions," *Applied Financial Economics*, 14, 937-943.

Ioannidis, J. P. A. (2005), "Why Most Published Research Findings Are False," *PLoS Medicine* 2 (8): e124

Jennions, M.D. and Møller, A.P., "Relationships fade with time: a meta-analysis of temporal trends in publication in ecology and evolution," *Proceedings of the Royal Society*, published online 4 December 2001; https://researchers.anu.edu.au/publications/8228

Kahneman, Daniel, and Amos Tversky, 1971, "Belief in the law of small numbers," *Psychological Bulletin*, 76 (2), 105-110.

Kahneman, Daniel, A. Tversky. 1972. "Subjective probability: A judgment of representativeness," *Cognitive Psychology*, 3 430-454.

Kahneman, Daniel, and A. Tversky, 1973, "On the psychology of prediction," *Psychological Review*, 80, 237–251.

Kahneman, Daniel, and A. Tversky, 1979, "Prospect theory: An analysis of decision under risk,"
Econometrica 47 (2) 263-292.

Keil, M., G. Smith, and M.H. Smith, 2004. "Shrunken Earnings Predictions are Better Predictions," *Applied Financial Economics*, 14, 937–943.

Kelley, T. L. 1947. *Fundamentals of Statistics*. Cambridge, MA: Harvard University.

Kim, M., Nelson, C., and R. Startz, 1991, "Mean Reversion in Stock Prices? A Reappraisal of the Empirical Evidence," *The Review of Economic Studies*, 58 (3), 515-528.

Kumar, Alok. 2009, "Who gambles in the stock market?" *Journal of Finance*, 64 (4), 1889-1933.

Kuskowska-Wolk A., Karlsson P, Stolt M, Rossner S, 1989, "The predictive value of body mass index based on reported weight and height," *International Journal of Obesity*, 13, 441-443.

Khurana, Rakesh, 2002, "The Curse of the Superstar XEO," *Harvard Business Review*, 80 (9), 60-66.

Khurana, Rakesh, 2002, *Searching for a Corporate Savior: The Irrational Quest for Charismatic CEOs*, Princeton, NJ: Princeton University Press.

King, W. I., 1934. "Review of The Triumph of Mediocrity in Business by Horace Secrist," *Journal of Political Economy*, 42 (3), 398–400.

Kirkley, Alexandra, Trevor B. Birmingham, Robert B. Litchfield, J. Robert Giffin, Kevin R. Willits, Cindy J. Wong, Brian G. Feagan, Allan Donner, Sharon H. Griffin, Linda M. D'Ascanio, Janet E. Pope, and Peter J. Fowler, "A Randomized Trial of Arthroscopic Surgery for Osteoarthritis of the Knee," *New England Journal of Medicine*, September 11, 2008; 359:1097-1107.

La Porta, R., 1996, "Expectations and the cross-section of stock returns," *Journal of Finance*, 51 (5) 1715–1742.

Lakonishok, J., A. Shliefer, R. W. Vishny, 1994, "Contrarian investment, extrapolation, and risk," *Journal of Finance*, 49 (5) 1541–1578.

Langer, E. J. and J. Roth, 1975, "Heads I win, tails it's chance: The illusion of control as a function of the sequence of outcomes in a purely chance task," *Journal of Personality and Social Psychology*, 32 (6) 951-955.

Laughhunn, D. J., J. W. Payne, 1980, "Translation of gambles and aspiration level effects in risky choice behavior," *Management Science*, 26 (10) 1039-1060.

Lee, Marcus, and Gary Smith, 2002, "Regression to the Mean and Football Wagers," with Marcus Lee, *Journal of Behavioral Decision Making*, 15 (4), 329–342.

Lehrer, Jonah, "The truth wears off," *The New Yorker*, December 13, 2010.

Lev, Baruch, 1969, Industry Averages as Targets for Financial Ratios, *Journal of Accounting Research*, 7 (2), 290-299.

Li J, Zhang Q, Zhang M, Egger M, "Intravenous magnesium for acute myocardial infarction," Cochrane Evidence, January 21, 2009.

Li J, Zhang Q, Zhang M, Egger M. "Intravenous magnesium for acute myocardial infarction," Cochrane Database of Systematic Reviews, 2007, Issue 2.

Light, R. J., D. B. Pillemer (1984), *Summing up: The Science of Reviewing Research*, Cambridge, Massachusetts: Harvard University Press.

Limbaugh, Rush, radio transcript, "No, I'm Not Moving to Costa Rica," March 09, 2010.

Locke, P. R., S. C. Mann. 2004, "Prior Outcomes and Risky Choices by Professional Traders," Working Paper, Texas Christian University.

López-Abente G, A Escolar, 2001, "Tobacco consumption and bladder cancer in non-coffee drinkers," *Journal of Epidemiology and Community Health*, 55: 68-70.

Lord, F. M., and M.R. Novick. 1968, *Statistical Theory of Mental Test Scores*, Reading, MA: Addison-Wesley.

MacMahon, B, Yen S, Trichopoulos D, et al, 1981, "Coffee and Cancer of the Pancreas," *New England Journal of Medicine*, 304: 630-633.

Marino PL (2007). "Antimicrobial therapy," *The ICU Book*, Hagerstown, MD: Lippincott Williams & Wilkins

Markusen, B, 2009, "The legend of Danny Goodwin," Hardball Times, retrieved September 15, 2015, from http://www.hardballtimes.com /the-legend-of-danny-goodwin/

Massey, Cade, and Richard H. Thaler, 2013, "The Loser's Curse: Decision Making & Market Efficiency in the National Football League Draft," Management Science, 59 (7), 1479-1495.

McQuaid, Clement, ed., 1971, *Gambler's Digest*, Northfield, Illinois: Digest Books, 24-25.

McQueen, G., 1992, "Long-horizon mean-reverting stock prices revisited," *Journal of Financial and Quantitative Analysis*, 27, 1–18.

McQueen, G., and S. Thorley, 1999, "Mining fool's gold," *Financial Analysts Journal*, 55 (2), 61-72.

Miao, L. L., "Gastric Freezing: An Example of the Evaluation of Medical Therapy by Randomized Clinical Trials," in J. P. Bunker, B. A. Barnes, and F. Mosteller, editors, *Costs, Risks and Benefits of Surgery*, New York: Oxford University Press, 1977, pp. 198–211.

Morton, Veronica, David J Torgerson, 2003, "Effect of regression to the mean on decision making in health care," *British Medical Journal*, 326: 1083-1084.

Moseley, J. Bruce, Kimberly O'Malley, Nancy J. Petersen, Terri J. Menke, Baruch A. Brody, David H. Kuykendall, John C. Hollingsworth, Carol M. Ashton, and Nelda P. Wray, "A Controlled Trial of Arthroscopic Surgery for Osteoarthritis of the Knee," *New England Journal of Medicine*, July 11, 2002; 347:81-88.

Newhan, Ross, 1989, "Dodgers Continue Downfall," *Los Angeles Times*, May 14.

Neisser, Ulric, Boodoo, Gwyneth, Bouchard, Thomas J. Jr., Boykin, A. Wade, Brody, Nathan, Ceci, Stephen J., Halpern, Diane F., Loehlin, John C., Perloff, Robert, Sternberg, Robert J., Urbina, Susana, 1996, "Intelligence: Knowns and unknowns," *American Psychologist*, 51 (2), 77-101.

Nelson, K. 2003, "What poker can teach you about investing," Legg Mason Funds Management

Investment Conference.

Niendorf, B., and K. Beck, 2008, "Good to great, or just good?," *Academy of Management Perspectives*, 22 (4), 13-20.

Palfreman, Jon "The Rise and Fall of Power Line EMFs: The Anatomy of a Magnetic Controversy," *Review of Policy Research*. 23 (2), pages 453–472, March 2006.

Peter, Laurence J., and Hull, Raymond, 1969, *The Peter Principle: Why Things Always Go Wrong*, New York: William Morrow and Company.

Peters, T., and Waterman, R.H. Jr. 1982. *In Search of Excellence*, New York, Harper & Row.

Poe, Edgar Allan, 1845, "The Mystery of Marie Roget," *Tales*, 151-199.

Poterba, J. M. & Summers, L. H., 1988, "Mean reversion in stock prices: Evidence and implications," *Journal of Financial Economics*, 22 (1), 27–59.

Pritchett, Lant, and Lawrence H. Summers, 2014, "Asiaphoria Meets Regression to the Mean," *National Bureau of Economic Research*, Working Paper 20573.

Quirk, Chris, "Herd Mentality," *Harvey Mudd College Magazine*, 25-27.

Resnick, B. G., and T. L. Smunt, 2008, "From good to great to . . ." *Academy of Management Perspectives*, 22 (4), 6-12.

Richardson, M., 1993, "Temporary Components of Stock Prices: A Skeptic's View," *Journal of Business & Economic Statistics*, 11 (2), 199-207.

Riegel, R. 1933. "Review of The Triumph of Mediocrity in Business by Horace Secrist," *Annals of the American Academy of Political and Social Science*, 170, 178–179.

Roper Center for Public Opinion Research at Cornell University kindly provided the presidential approval data.

Rose G., Blackburn H., Keys A., et al. "Colon cancer and blood-cholesterol." *Lancet*, 1974; 7850: 181–183.

Rose, G. and M. J. Shipley, "Plasma lipids and mortality: a source of error," *Lancet*, 1980; 8167: 523–526.

Rousseeuw PJ. 1991, "Why the wrong papers get published," *Chance*, 4:41-3.

Rulon, Philip J. 1941, "Problems of Regression," Harvard Educational Review, 11 (2), 213-223.

Savitz, David A., Pearce, Neil E., and Charles Poole, 1989. "Methodological Issues in the Epidemiology of Electromagnetic Fields and Cancer," *Epidemiological Reviews*, 11, 59-78.

Schall, Teddy, and Gary Smith, 2000, "Do Baseball Players Regress toward the Mean?" *The American Statistician*, 54 (4), 231-235 (also 1999, "Proceedings of the Section on Statistics in Sports," *American Statistical Association*, 2000, 8–13)

Secrist, H. 1933, *The Triumph of Mediocrity in Business*, Evanston, Ill: Northwestern University, 1933.

Schlichtung P., Hoilund-Carlsen, Quaade F. 1981, "Comparison of self reported height and weight with controlled height and weight in women and men," *International Journal of Obesity*, 5, 67-76.

Sharpe, W.F. 1985, *Investments*, 3rd edition, Englewood Cliffs, New Jersey: Prentice-Hall, 430.

Sheard, R., 1997, "The Daily Dow," retrieved August 28, 2013, from http://www.fool.com/DDow/1997/DDow971230.htm.

Shiller, R.J. (1981) "Do Stock Prices Move Too Much To Be Justified by Subsequent Changes in Dividends?" *The American Economic Review*, Volume 71, June, No. 3 pages 421-435.

Shiller, R. J., 1984, "Stock prices and social dynamics," Brookings Papers on Economic Activity, 2, 457-498.

Smith, Gary, 1997, "Do Statistics Test Scores Regress Toward the Mean?," *Chance*, 42–45.

Smith, Gary, and Joanna Smith, 2005, "Regression to the Mean in Average Test Scores," *Educational Assessment*, 10, 377-399.

Smith, Gary, Joseph Steinberg, and Robert Wertheimer, 2006, "The Next Best Thing to Knowing Someone Who is Usually Right," *Journal of Wealth Management*, 9 (3), 51-60.

Smith, M. H., Keil, M., and G. Smith, 2004, "Shrunken earnings predictions are better predictions," *Applied Financial Economics*, 14, 937-943.

Smith, Gary, Levere, Michael Levere and Robert Kurtzman, 2009, "Poker Player Behavior After Big Wins and Big Losses," *Management Science*, 55 (9), 1547-1555.

Stanton, Jeffrey M., 2001, "Galton, Pearson, and the Peas: A Brief History of Linear Regression for Statistics Instructors," *Journal of Statistics Education* Volume 9, Number 3.

Steenbarger, B. 2007. "What we can learn from trading and poker," http://traderfeed.blogspot.com/2007/04/what-we-can-learn-from -trading-and.html

Stewart, Matthew, 2006, "The Management Myth," *The Atlantic*, June, 297 (5), 80-87.

Summers, L. H., 1986, "Does the stock market rationality reflect fundamental values?," *Journal of Finance* 41, 591–601.

Teo, Koon K, and Salim Yusuf, 1993, "Role of magnesium in reducing mortality in acute myocardial infarction. A review of the evidence," Drugs, 46 (3), 347-359.

Thaler, R., E. J. Johnson. 1990, "Gambling with the house money and trying to break even: The effects of prior outcomes on risky choice," *Management Science* 36 (6) 643-660.

Thorndike, Robert L., 1942, "Regression Fallacies in the Matched Group Experiment," *Psychometrika*, 7 (2), 85-102.

Thorndike, R. L., 1963, *The Concepts of Over- and Under-Achievement*, New York: Teacher's College, Columbia University.

Tversky, A., and D. Kahneman, 1973, "On the Psychology of Prediction," *Psychological Review*, 80, 237–251.

Tversky, A., and D. Kahneman, 1974, "Judgement Under Uncertainty: Heuristics and Biases," *Science*, 185, 1124–31.

Tversky, A., D. Kahneman, 1981, "The framing of decisions and the psychology of
Choice," *Science*, 211 (4481) 453-458.

Tversky, A., D. Kahneman, 1992, "Advances in prospect theory: Cumulative representation of uncertainty," *Journal of Risk and Uncertainty*, 5 (4) 297-323.

UK Childhood Cancer Study Investigators, "Exposure to power-frequency magnetic fields and the risk of childhood cancer," *Lancet*, 1999 Dec. 4; 354(9194): 1925-31.

Vinciguerra, Thomas, "Promises, Promises," *New York Times*, June 30, 2012.

Viscoli, Catherine M., Mark S. Lachs, and Ralph I. Horowitz, "Bladder Cancer and coffee Drinking: A summary of case-control research," *The Lancet*, June 5, 1993, 1432-1437.

Wagenmakers, EJ; Wetzels, R, Borsboom, D, van der Maas, HL (March 2011), "Why psychologists must change the way they analyze their data: the case of psi: comment on Bem (2011)." *Journal of Personality and Social Psychology* 100 (3): 426-32.

Wainer, Howard, 1999, "Is the Akebono School failing its best students? A Hawaii adventure in regression," *Educational Measurement: Issues and Practice*, 18 (3), 26–31, 35.

Wainer, Howard, 2007, "The Most Dangerous Equation," *American Scientist*, 95 (3), 249-256.

Wainer, Howard, and Lisa M. Brown, 2007, "Three Statistical Paradoxes in the Interpretation of Group Differences: Illustrated with Medical School Admission and Licensing Data," *Handbook of Statistics*, 26, 893-918.

Wallis, W. Allen, and Harry V. Roberts, *Statistics: A New Approach*, Glencoe Illinois: Free Press, 1956, 479-480.

Wertheimer, Nancy, and Ed Leeper, "Electrical Wiring Configurations and Childhood Cancer," *American Journal of Epidemiology*, 109 (3), 273-284.

Williamson, J.G. 1991, "Productivity and American leadership: A Review Article," *Journal of Economic Literature*, 29 (1), 51–68.

Zweig, J. 2009. "As stock losses loom, don't throw a 'Hail Mary'," *The Wall Street Journal*, February 21, 2009.

INDEX

Index

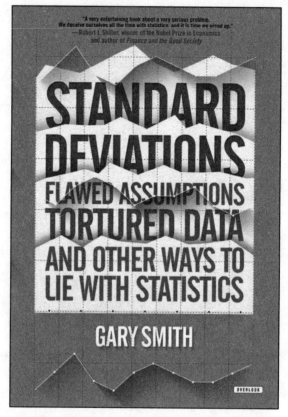